Applied Geography: Issues, Questions, and Concerns

The GeoJournal Library

Volume 15

Series Editor: WOLF TIETZE, Helmstedt, FR Germany

The titles published in this series are listed at the end of this volume.

Applied Geography: Issues, Questions, and Concerns

Edited by

MARTIN S. KENZER

Department of Geography and Anthropology
Louisiana State University
Baton Rouge, Louisiana
U.S.A.

(Currently Visiting Professor at the
University of Southern California)

KLUWER ACADEMIC PUBLISHERS
DORDRECHT / BOSTON / LONDON

Library of Congress Cataloging-in-Publication Data

Applied geography : issues, questions, and concerns / edited by Martin
S. Kenzer.
     p.   cm. -- (GeoJournal library ; 15)

   ISBN-13: 978-94-010-6697-6   e-ISBN-13: 978-94-009-0471-2
   DOI: 10.1007/978-94-009-0471-2

   1. Geography--Philosophy.   I. Kenzer, Martin S., 1950-
II. Series: GeoJournal library ; v. 15.
G70.A815   1989
910'.01--dc20   89-19740

Published by Kluwer Academic Publishers,
P.O. Box 17, 3300 AA Dordrecht, The Netherlands.

Kluwer Academic Publishers incorporates
the publishing programmes of
D. Reidel, Martinus Nijhoff, Dr W. Junk and MTP Press.

Sold and distributed in the U.S.A. and Canada
by Kluwer Academic Publishers,
101 Philip Drive, Norwell, MA 02061, U.S.A.

In all other countries, sold and distributed
by Kluwer Academic Publishers Group,
P.O. Box 322, 3300 AH Dordrecht, The Netherlands.

Printed on acid-free paper

To Dagney Anne Kenzer McKinney

Table of contents

SECTION IV: APPRAISALS FROM HUMAN GEOGRAPHERS

SECTION V: THE 'TAKEN FOR GRANTED' SIDE OF APPLIED GEOGRAPHY

Preface

The completion of this collection took many months, and, for a variety of reason, required the assistance and/or indulgence of a number of individuals. First and foremost, I would like to thank Tim Hudson for his useful input and support at the outset of the project. Likewise, I would like to thank Jesse O. McKee for providing a hospitable environment during my affiliation with the University of Southern Mississippi.

At Louisiana State University I am grateful to Sam Hilliard and Carville Earle for their invaluable understanding. The book became part of the GeoJournal Library as a result of Wolf Tietze's confidence in the topic, and because of Henri G. van Dorssen's (and Kluwer Academic Publishers') good nature – despite numerous 'problems'. Curtis C. Roseman, and the remainder of the Geography Department at the University of Southern California (where I completed many last-minute details for the volume), are to be thanked for the cordial and warm environment I received while a visitor in Los Angeles.

Finally, no multi-authored collection reaches completion without the help of many patient contributors. This particular book suffered many set-backs along the way, so I am particularly grateful to the authors herein. They demonstrated their compassion and exceptional professionalism throughout, by never second-guessing my decisions, and by allowing me to remedy the set-backs in my own way. They were a pleasure to work with, and they should take pride in their achievements.

Los Angeles, California

M. S. Kenzer (ed.), Applied Geography: Issues, Questions, and Concerns, ix.
© 1989 *Kluwer Academic Publishers.*

Applied Geography: Overview and Introduction

For well over a decade, many professional geographers in the United States have demonstrated considerable interest in the applied aspects of their discipline. Their attention has been focused on job opportunities rather than geography's traditional role as a problem-solving discipline that contributes to a broad, well-rounded, liberal arts education. A similar pattern developed later in Canada, and it spread to England shortly thereafter. The 'applied movement' is today well ensconced in academic geography programs in many countries.

The movement arose a decade ago in response to the 'sterile' quantifying geography of the 1960s and early 1970s. 'Relevancy' was the key word in 1975, and the motivation behind applied geography was to make geographic research relevant to social needs and policy formulation (see Sant 1982). A spate of articles on 'relevant geography' appeared in the literature, followed by a call for the restructuring of the discipline to serve the public's needs. It was pointed out that the 'ambiguous' term 'applied geography' dates back to the mid-1950s (Harrison 1977; cf. Dunbar 1978), but that this new 'applied geography' would be both relevant and rigorous, serving the concerns of the private and public sectors, and utilize *scientific* methods. The call was to make the discipline marketable and viable (Harrison and Larson 1977; Moriarty 1978), to alter geography's theretofore image as a field for 'teaching-people-to-teach-people-to-teach-people' (see Beard 1976), and to broaden its base of influence.

Concomitant with the relevancy issue was a marked decline in the number of students throughout North America. The post-World War II, baby-boom generation had reached college age, and student enrollment was predictably dropping across all disciplines. To identify and cultivate as many potential practitioners as possible – realizing that it was self-defeating to train students for teaching positions alone – the outcry was to adopt a long-term policy that would help graduates find jobs outside of academia (Marcus 1978; Wilbanks and Libbee 1979; Mikesell 1979, 1980; Mayhew 1980; Frazier 1980; Stutz 1980). 'Relevancy' and 'curriculum change' were regarded as complementary endeavors to facilitate the applied movement (Frazier 1978).

Publication outlets for applied research and news of opportunities for applied employment appeared almost overnight: the *Geographical Review* announced the inclusion of a section on applied research (1976); a national survey noted that nearly 90 percent of the sampled geography departments, as early as 1977, were willing to place greater emphasis on applied geography over the next five years (Harrison 1977); the *Professional Geographer* and the *AAG Newsletter* undertook to reorient their respective formats to include sections on applied geography; an 'applied geography conference,' now an annual event, first convened in 1978; the

M. S. Kenzer (ed.), Applied Geography: Issues, Questions, and Concerns, 1–11.
© 1989 *Kluwer Academic Publishers.*

international journal *Applied Geography* appeared in 1980; the *Applied Geography Newsletter* began publication in 1980; a *Directory of Applied Geographers* (available from the AAG, Association of American Geographers) surfaced in 1981; and the movement diffused to Canada where the *Operational Geographer* was born in 1983. All indications suggest that a parallel situation has emerged in England (Johnston 1986).

Although there were a few precautionary notes warning of the dangers of an 'unchecked' applied movement (Ford 1982; Salter 1983; Hornbeck 1979), and whereas a handful of practitioners tried to call attention to the contradictory purposes of an applied geography on the one hand and an academic geography emphasizing core, traditional specialties on the other (Munski 1984; McDonald 1981; Hill 1981; Grosvenor 1984; Hart 1982), the reactions to these ostensibly slight murmurs of protest were significant and immediate (e.g., Frazier 1982; Moorlag 1983; Smith and Hiltner 1983; Russell 1983). Recent dialogue, however, has centered on the potential collaborative efforts of academia, the government, and the private sector (e.g., Marotz 1983; Monte 1983; Lier 1984; Bradbury 1985). Smith and Hiltner (1983), moreover, have shown that applied geography has become more widespread, and that the training of applied geographers tends to reflect the same regional changes taking place within the discipline in America as a whole (i.e., from the Far West and the Midwest to the Great Lakes and East Coast regions). But, critical debate is still nonexistent.

One result of extra effort having been placed upon the applied opportunities in geography has been an increased awareness that traditional subdisciplinary specializations, such as cultural, political, historical, and regional geography, are rapidly declining in prominence, while new academic appointments are being filled primarily in the 'techniques' sector (Green 1983; Hausladen and Wyckoff 1985). One need only glance through the recent 'Jobs in Geography' section of the *AAG Newsletter* to see the emphasis American geography departments are attributing to geographic information systems (GIS) and other technique-oriented skills. A couple of years ago, GIS was not recognized by the AAG as a specialty; by 1987 it constituted the third largest specialty group of the association; one year later it was nine members short of becoming the association's largest specialty group (AAG 1989b). Beginning September, 1989, a new National Center for Geographic Information and Analysis (federally funded by the National Science Foundation, see AAG 1988) was offering a 'model GIS curriculum' for adoption by as many colleges and universities as possible (AAG 1989a). Fortunately, at least one geographer has forewarned his colleagues of the strong competition they are up against when competing for intellectual respectability in the 'new' 'field' of GIS (Douglas 1988).

To facilitate an applied stance, growing numbers of geographers have developed increasingly narrow specialties, to the extent that others fear the discipline has lost touch with its long-term synthetic heritage – i.e., explicating the relationship between people, places, cultures, and the global/regional mix of each (see Lewis 1985). As more and more emphasis has been placed on techniques and their application to everyday geographical concerns, a concomitant tendency (at least among human geographers) has been a marked decline in real world, field-oriented experience (Rundstrom and Kenzer 1989); the application of GIS and related technology requires little or no fieldwork. Synthesis and fieldwork have always formed a major segment of geography's traditional core, yet that core is potentially due to vanish (as are the geographers capable of contributing to it) should applied geography continue to gain dominance over the academic side of the field. In sum, the academic side of geography may be losing its very foundations as more and more students are trained *not* to think critically, but, instead, to serve as technicians and address the needs of business and government. Wynn has lamented on this trend occurring throughout the English-speaking world:

> Of all the recognized university disciplines, geography offers perhaps the greatest scope for realizing the goals of liberal education within a single department ... geography should enhance an understanding of the complex diversity of man's habitat, promote a sense of cultural relativism, provide a rich context for understanding current and local events, foster a critical spirit, encourage the testing of generalizations (abstractions) against experience, and develop an appreciation of man's role in shaping the earth ... To abandon this goal in an attempt to tailor our graduates to the specific skill requirements of employers is to abrogate our responsibilities as *educators* and to undermine one of the central attractions of our discipline as a field of inquiry (emphasis in original) (Wynn 1983, p. 66).

Regrettably, the current spate of applied geography research (at least in North America) does not seem to have any philosophical or theoretical basis in being, other than its understood application to social needs (see Kenzer 1984). In some cases, especially in the United States, it appears to stem solely from the public's vague perception of what geographers do, combined with a subsequent reduction of federal and state dollars to support geographic research. Compounding the problem, only a minuscule percentage of all undergraduate students elect geography as their declared major (Harrison and Larson 1977; Wilbanks and Libbee 1979), causing some practitioners to fear that belt-tightening measures may adversely affect geography in a disproportionate, if not severe manner. Diverting students away from careers in academia and in to more applied, nonacademic specializations has been the proposed solution. A recent past-president of the AAG, in fact, proclaimed that 'the employment and research frontier for geography lies in the non-academic sectors of the economy' (Demko 1988, p. 575). Yes, this creates jobs, but what of the discipline's intellectual core? Hence, while the applied movement is arguably a stop-gap measure, and whereas its adherents have been pragmatically motivated, they have given seemingly little consideration to its nature and long-term effect on the discipline's academic standing.

On the other side of the fence, there are geographers who do *not* consider themselves 'applied geographers,' believing, instead, in the inherent value of 'pure' or 'theoretical' research. In their view, academic research represents the frontier of geography, and those geographers who work outside of academia rely on academic insights and then apply them to the continual problems we face in an imperfect world. Accordingly, this group questions the validity, vigor, and usefulness of an *academic* discipline that overly emphasizes applied research and/or public promotion campaigns. As academic practitioners first and foremost, they are justifiably concerned about the long-term implications of applied research, especially if it results in the near exclusion of all other geographic endeavors (see Jordan 1988a, 1988b). There does not, however, appear to be a single definition of what constitutes applied geography, nor how, and to what extent, it differs from so-called pure geography (Dunbar 1978; Ford 1982; Kenzer 1984.) Even those working outside of academia have questioned both the practice and the desirability of classifying practitioners as either applied or nonapplied geographers, as if they were mutually exclusive avocations (Richardson 1989).

Despite informal complaints voiced at professional meetings and a growing concern noted in personal correspondence, however, critical discourse on the purpose and nature of applied geography has failed to materialize. Indeed, key ethical questions regarding the appropriateness of the applied geographers' usurping of the disciplinary mantle remain unasked. Applied geography (of an entirely different variety) was twice before of major concern to the discipline: during and between the two world wars. Why again now? Applied geographers rarely address that question, thereby adding to the overall confusion surrounding the present rise of the applied movement.

The purpose of this collection is to focus rigorous attention on the applied movement in geography: to determine whether a common, working definition of applied geography exists; to understand the rationale behind the applied movement; to uncover the degree to which applied geography has become a major research priority within the discipline; and to document curricula changes in geography departments as a result of the applied movement. The objective is an explication and better understanding of the movement's origins, its current and anticipated influence, the arguments both for and against it, and the extent to which an academic geography affected by the applied movement will be able to satisfy its societal obligation to address the public's trust (see Chouinard and Kenzer 1982).

It has been opined that 'Applied research in the public sector is driven by short-term, often highly political, considerations. The research is often of marginal quality due to time pressures, inadequate data bases, poor analytic procedures, and a lack of appropriate modeling protocol' (Haynes 1983, p. 8). However, this is mostly conjecture. While it is a fact that individual research endeavors, project proposals, funding criteria, newsletter announcements, journal articles, conference themes, disciplinary emphases, job announcements, department offerings, and associations' 'directives' all acknowledge, even advocate, the applied movement, research on the movement itself is spotty. Seemingly, applied geography is being regarded and promoted *not* as one of many subfields within the discipline, but rather as *the* research frontier for the whole of the field – hence, we need not spend time studying it! As a result, little has been written on the nature of applied geography; far less has been written questioning its contributions to the discipline. This writer knows of *no* published works examining the long-term influence of the applied movement on the discipline as a whole.

I initially conceived of this collection *not* as an indictment of applied geography

– since, ideally, all disciplines necessitate a balance between theoretical and applied research, whereby one half complements the other (Burnett 1977; Kenzer 1989a) – but, rather, to examine the possible implications and long-range results of a rapidly growing trend before the trend took permanent root. I asked the contributors herein to take opposing views: some were supposed to write about applied geography from a positive perspective – to discuss its inherent value and ultimate contributions to the discipline – while others were asked to approach it from a more critical vantage point – to suggest possible implications or potential side effects in the long run. The point was to stimulate discussion. In the end, many, but not all, of the contributors in this collection found both optimistic (good) and pessimistic (bad) things to say; several took a more middle-of-the-road stand, attesting to the reciprocal nature of pure and applied research. Each contributor was selected on the basis of his or her subdisciplinary interest. I tried to identify as diverse a group of geographers as possible, while still hoping to have something substantive in the end, and thus compare one essay (or one group of essays) with another.

Beyond this introductory chapter, the book is divided into five sections; each section explores broad interests and/or subspecialties within geography. The first section, 'Ruminations and Philosophical Queries,' contains two introspective essays on the nature and purpose of applied geography. John Fraser Hart begins the discussion by asking 'Why Applied Geography?' For him, all geography is applicable, so why bother adding the adjective 'applied' in front of 'geography'? According to Hart, a geographical perspective helps people understand the world around them, which is the essential criterion, whether or not a practitioner opts to attach the nametag 'applied' onto his or her business card. Professional geographers in all sectors of the economy add to the public's image of the field, so the critical caveat is to do as good a job as possible, irrespective of one's title. Hart ends with a plea for academic and nonacademic geographers to work toward a common end and not bicker over petty differences of opinion. In the second essay, Stubbings and Haynes provide three homilies – which they define as 'methodological and philosophical cautions' – for professional geographers: 1) thin conceptualization; 2) deflationary ethics; and 3) intellectual self-immolation. The first homily refutes the notion that applied research is value-free, or that geographers working in the private or public sector are nonpolitical; there is no such thing as neutral work or research, and applied geographers must constantly remind themselves of this point. Homily number two is directly related to the first. In sum, the call is for a more moral geography, recognizing that our actions are accountable for many short- and long-term events, whether or not we are cognizant of such events; applied geographers, in essence, are policy advisors and must keep this fact in mind. Finally, the third homily implies that self-promotion, in the form of research dollars and government/institutional priorities, often supersedes academic integrity. We sometimes let the available monies dictate our agendas, even when we lack interest in the projects being funded. The warning is that applied geographers, for practical purposes, may indulge themselves solely in areas where there are grants, thereby ignoring the more socially responsible research topics.

The following section, 'The Education Issue,' examines the impact and influence of applied research on the teaching side of the discipline. In the initial paper, Boehm and Harrison argue the virtues of an applied curricula, using the Southwest Texas State University (SWT) Department of Geography as their model. They define applied geography as 'the use of geographic content, principles, and methods in research and other activities designed to aid in the resolution of human problems...,' and they discuss ways in which the program in their Department trains students to implement that definition into internship programs and, later, into their professional careers. The SWT model seemingly works. The Department began a Masters of Applied Geography degree in 1983, and it can probably boast today of the largest applied geography program anywhere. Numerous geography programs and departments elsewhere have looked for guidance in SWT's success – and as a feasible model to emulate. Bruce Smith's paper, the second one in this section, looks at applied geography and experiential education (cf. Foster and Jones 1977). Like the previous paper, Smith uses his own university as an example, focusing on the substantial links between the Cooperative Education Program and the Geography Department at Bowling Green State University. The fundamental assertion is that on-the-job-training provides a critical and needed background to the students' repertoire of skills – irrespective of whether they enter the academic or the nonacademic job market. Professional geographers must learn to compete better with other disciplines (see Kenzer 1989b), and an internship program is certainly a step in the right direction. But, as Smith correctly notes, experiential education programs also have decided drawbacks, some of which apply to the student, while others pertain to faculty. The last education essay is Victoria Rivizzigno's inspection of recent, introductory college-level human geography textbooks. Basically, she contends that geography texts during the 1970s were far more analytical and theoretical in nature than the texts of the 1980s, which presently tend to lean more heavily on description. As she notes, this change in approach mirrors the larger disciplinary changes affecting the field as a whole. Unfortunately, her survey is limited to the U.S. textbook market, so a natural extension of this inquiry would be to peruse recent texts in other countries for contrasts and comparisons. The educational milieu is an important one, for this is the arena in which ideas are passed on explicitly to students. Some of these students then become the future generation of practicing geographers.

The third section of the book, 'Considerations in Physical Geography,' contains three chapters that call attention to the relationship between applied and theoretical research by physical geographers. Applied physical geography probably predates all other applied specialties within the discipline, so it is instructive to read what physical geographers have learned from their fertile past. The first two authors in this section are climatologists, but there is considerable disagreement in their respective attitudes toward applied research. John Lier's rather philosophical essay argues that geographical pursuits, throughout human time, have been involved with real world problems, so, at bottom, all geography is applied to one degree or another. He notes that many who call themselves applied geographers today are involved in off-campus consulting or research, but that their techniques and findings are meaningless without a solid academic training. A large segment of his paper demonstrates that to discriminate between applied and nonapplied is a useless exercise, at least in terms of climatological research. The only valid distinction is between those who engage in purely applied research, to the exclusion of all academic endeavors, and thus fail to add to the pool of knowledge and consequently rob the profession of valuable research findings. Glen Marotz, on the other hand, is a climatologist who sees things differently. For Marotz the applied route is the direct route: to apply the results of climatological and meteorological findings to human and environmental problems is crucial in his view – and the sooner the better. His preference is not to wait for those who *might* apply their education to real world issues, but, instead, to train students with a keen appreciation for applications from the start. However, as he ultimately suggests, even though there is less controversy in this subfield than in others (since climatology has always had a foot in both doors), the status and growth of applied climatology is to some extent unclear. The last chapter in this section is by Douglas Sherman, a coastal geomorphologist. Geomorphology has been one of the most applied-oriented subfields in the discipline, and this is as true today as in the past. Indeed, Sherman's broad working definition of applied geomorphology – 'using geomorphological theories and methods to address environmental problems' – certainly would lend itself to include a vast number of published geomorphology papers and monographs. He delineates five reasons for the current 'explosion' in applied geomorphology: the writings and influence of A.N. Strahler, the rise of process geomorphology, a concern for ecological issues, geomorphology as an aid to urban environmental factors, and the human impact on the physical landscape, or what Sherman calls 'anthropogeomorphology.' Applied and pure geomorphology complement one another, but, according to Sherman, those undertaking applied work need to remind themselves of their obligation for objective research, and not compromise their academic training; good science, he reminds us, generates good results, whether theoretical or applied.

The penultimate section, 'Appraisals from Human Geographers,' is comprised of three chapters by four human geographers who conduct very different types of geography. The first paper is by Daniel Gade, a cultural geographer. Gade sets out to show why cultural geographers have rarely exhibited a predilection for applied research. He attributes the existence of the subfield in the United States to Carl Sauer who, unlike many of his contemporaries, found a historical approach urgent to geographical inquiry. The dimension of time forces one to view data not as an end in and of themselves, but as part of some amorphous, unfolding continuum. Likewise, a joint emphasis on cultural artifacts – whether it be the entire human landscape or portions of material culture – stresses place over space and explanations gravitate toward the probable rather than the exact. For these reasons, cultural geographers have been less willing to study 'objective' 'facts,' preferring, instead, immediate experience. Gade does, however, give ample instances where cultural geographers have done useful, decision- and problem-oriented research, concluding that while this approach is beneficial, an academic discipline should never put undue weight on applied, contract-based efforts. The second paper in this section is an overview of 'Applied Recreation Geography' by Janiskee and Mitchell.

Recreation geography is among the more recent subspecialties, as well as one of the most application-oriented subfields in the discipline; Janiskee and Mitchell define it as 'geographic expertise used as a tool for solving leisure-related problems.' The authors indicate that the vast majority of research endeavors under this label typically concentrate on leisure activities of some sort (recreation, tourism, and sport), and, in fact, they posit a possible change of name to 'leisure geography.' They suggest that the applied versus pure debate has little import in recreation geography, but believe, instead, that the more relevant question is whether recreation geographers will throw their support and expertise behind predictive and normative approaches toward problem solving research, and to do so on a more widespread basis. In the end they argue that both sides of the debate are weakened when applied practitioners lack a rigorous, scientific orientation. David Hornbeck's essay concludes the section on human geography. Hornbeck is both a respected historical geographer and an economic geographer who owns a success-ful consulting firm. Thus, as the title of his paper implies, he knows 'both sides of the street,' and is able to provide a balanced perspective on the virtues of both applied *and* nonapplied geography. In his view, applied and academic geography are virtually identical, with little or no difference in their end product. Hornbeck thus sees no reason to alter existing curricula further, as has been the proposed remedy (or has been enacted in certain cases throughout North America). His point is that the job market is in constant flux, so it makes little sense to try to adapt to every market whim; again, high quality work engenders positive results and demands respect. The key, he writes, is *not* to develop new job-related courses, but to improve the courses we now offer and 'include more breadth in the discipline and emphasize analytical and communication abilities.' Further, he opines that business and geography are two distinct ventures and that business (applied) geographers, to survive in the business world, must be adept at business first and foremost, at geography secondarily; this fully distinguishes them from academic geographers, who probably have little need to appreciate the business environment.

In the fifth and final section, 'The "Taken for Granted" Side of Applied Geo-graphy,' are two essays that treat aspects of geography many automatically assume are related to the applied side of the field. The first paper is about cartography – the subfield that too many professional geographers consider the handmaiden of the discipline – while the second is focused on the contributions of women in geo-graphy – those taken-for-granted practitioners who, until recently, rarely (if ever) received the credit they were due. Robert Rundstrom reminds us that maps serve a purpose because they are useful. This *use*ful component of maps lends itself to the popular notion that cartography and cartographers exist for the benefit of others, including other geographers. Many professional geographers believe that carto-graphers merely assemble facts into a pictorial form – i.e., while the 'true' geo-grapher goes directly to the source (the field) and collects the necessary research material, it is the cartographer who simply takes this data and constructs a map from them. Since there is theoretically no creativity involved, cartographers need not concern themselves with anything beyond the end product. This, basically, is the public's view as well, which adds to the notion that cartographers only follow orders. Rundstrom reviews the origins of this attitude and makes a plea to bring cartography back onto the academic side of the field – a position that ultimately gains greater respect for geographers and cartographers alike. As he rightly points out, the 'current priorities [of cartographers] are weighted too heavily toward development and training in applications of short-term value, almost to the exclusion of research and education as a long-term investment.' In the final chapter, Alice Andrews directs our attention toward the careers and contributions of five women who made their mark on the applied side of the discipline, even though contemporary geographers may know virtually nothing of their existence. The five women are Helen Strong, Clara LeGear, Evelyn Pruitt, Betty Didcoct Burrill, and Dorothy A. Muncy, each important to the advancement of the discipline in her own way. Andrews concludes that professional women geographers in the United States still occupy a minor role in the academic sector; she notes, however, that women represent over one-third of all geography majors. Further, the more applied avenues of employment – particularly in private industry, at research centers, at nonprofit organizations, and in the self-employed realm – continue to provide far greater opportunities for female geographers; federal employment, on the other hand, seems to have lost its prominence as the primary employer of professional women trained in geography (cf. Russell 1983). The key question is, 'what happens to women between the student and professor stage to account for their reduced numbers?' In other words, why are they perceived so positively by nonacademic employers, when they are seemingly viewed with less enthusiasm within academia?

The applied side of geography has always been a vital and important component of the discipline. In one sense, one might argue that *all* geography is applied, since the field is often depicted as a synthesizing discipline that borrows from other fields and then applies those findings to a so-called geographic problem. Hence, no author herein is advocating its eradication. Yet, passively, or at least inadvertently, and with little forethought to its long-range disciplinal effect, certain geographers have systematically promoted a business-related, profit-motivated applied geography for over a decade, which all too frequently has neglected geography's intellectual core. This has stimulated the numerous issues, questions, and concerns about the applied movement addressed in this collection.

On its present trajectory, the applied movement has fundamental consequences for professional geography in at least four areas: the evolution of the academic curricula; the traditional core research concerns of the field; the perception of the field by and our relations with other academic disciplines; and the employment opportunities beyond the halls of academia. As the following thirteen chapters attest, the time is indeed ripe to discuss constructively the direction the movement is taking the discipline, to evaluate critically the potential long-term effects of the movement on the academic side of the discipline, and to anticipate the eventual shortfall of geographers trained in the more traditional, theoretical areas of geographic research. Unfortunately, the division between applied and nonapplied is widening, and the bitter seeds of factionalism are beginning to appear (see Richardson 1989). It is the hope of this writer and of all the contributors to this collection that this book will stimulate wholesome dialogue and help mend fences. Neither the applied nor the academic side will benefit from prolonged internal conflicts and piddling arguments over 'relevant' geographic research.

Bibliography

AAG. 1988. National center for GIS to be launched with NSF grant. *AAG Newsletter* 23(8): 1.

AAG. 1989a. Testing of model GIS curriculum to begin in September. *AAG Newsletter* 24(3): 1.

AAG. 1989b. 1988 AAG topical and areal proficiencies and specialty group membership. *AAG Newsletter* 24(3): 10.

Beard, D.P. Professional problems of nonacademic geographers. *Professional Geographer* 28: 127–131.

Bradbury, J. 1985. Jobs, cogs and relevance in the 1980's. *Operational Geographer* 3(3): 4–5.

Burnett, G.W. 1977. Applied geographers and the marketability of geographers. *Professional Geographer* 29: 109–110.

Chouinard, V.A. and Kenzer, M.S. 1982. A question of hours or an hour of questions?: a response to 'the sixteen-million-hour question,' by Marvin W. Mikesell. *Professional Geographer* 34: 93–95.

Demko, G.J. 1988. Geography beyond the ivory tower. *Annals of the Association of American Geographers* 78: 575–579.

Douglas, D.H. 1988. Hardball and softball in geographic Information systems. *Operational Geographer* 6(3): 41–43.

Dunbar, G.S. 1978. What *was* applied geography? *Professional Geographer* 30: 238–239.

Ford, L.R. 1982. Beware of new geographies. *Professional Geographer* 34: 131–135.

Foster, L.T. and Jones, K.G. Applied geography: an educational alternative. *Professional Geographer* 29: 300–304.

Frazier, J.W. 1978. On the emergence of an applied geography. *Professional Geographer* 30: 233–237.

Frazier, J.W. 1980. Comments on Mayhew's 'new opportunities for the geographer'. *Professional Geographer* 32: 372–374.

Frazier, J.W. ed. 1982. *Applied Geography: Selected Perspectives.* New York: Prentice-Hall.

Green, D.B. 1983. Teaching positions in geography in the United States: what specialties have been in demand? *AAG Newsletter* 18: 14–15.

Grosvenor, G.M. 1984. The society and the discipline. *Professional Geographer* 36: 413–418.

Harrison, J.D. 1977. What *is* applied geography? *Professional Geographer* 29: 297–300.

Harrison, J.D. and Larson, R.D. 1977. Geography and planning: the need for an applied interface. *Professional Geographer* 29: 139–147.

Hart, J.F. 1982. The highest form of the geographer's art. *Annals of the Association of American Geographers* 72: 1–29.

Hausladen, G. and Wyckoff, W. 1985. Our discipline's demographic futures: retirements, vacancies, and appointment priorities. *Professional Geographer* 37: 339–343.

Haynes, K.E. 1983. Scientific geography, applied research, and public policy analysis. *Pittsburgh Geography and Regional Science Sessions Newsletters* 1: 7–8.

Hill, A.D. 1981. A survey of the global understanding of American college students: a report to geographers. *Professional Geographer* 33:237–245.

Hornbeck, D. 1978. Applied geography, is it really needed? *Journal of Geography* 78: 47–49.

Johnston, R.J. 1986. *On Human Geography*. New York: Basil Blackwell.

Jordan, T.G. 1988a. The intellectual core. *AAG Newsletter* 23(5): 1.

Jordan, T.G. 1988b. Anti-intellectualism. *AAG Newsletter* 23(6): 2.

Kenzer, M.S. 1984. Comments from the outside: the sixth annual applied geography conference, 12–15 October, 1983. *Applied Geography* 4: 85–86.

Kenzer, M.S. 1989a. Applied geography. In *Modern Geography: An Encyclopedic Survey*, ed. G.S. Dunbar, in press. New York: Garland Publishing Co.

Kenzer, M.S. ed. 1989b. *On Becoming a Professional Geographer*. Columbus, OH: Merrill.

Lewis, P. 1985. Beyond description. *Annals of the Association of American Geographers* 75: 465–478.

Lier, J. 1984. Comments on industry-government-academic cooperation. *Professional Geographer* 36: 219–221.

Marcus, M.G. 1978. The association of American geographers: planning for the future. *Professional Geographer* 30: 113–122.

Marotz, G.A. 1983. Industry-government-academic cooperation: possible benefits for geography. *Professional Geographer* 35: 407–415.

Mayhew, B.W. 1980. New opportunities for the geographer. *Professional Geographer* 32: 232–233.

McDonald, J.R. 1981. Revitalizing regional geography: an approach at the graduate level. *Journal of Geography* 80: 151–152.

Mikesell, M.W. 1979. Current status. *Professional Geographer* 31: 358–360.

Mikesell, M.W. 1980. The sixteen-million-hour question. *Professional Geographer* 32: 263–268.

Monte, J.A. 1983. The job market for geographers in private industry in the Washington DC area. *Professional Geographer* 35: 90–94.

Moorlag, S. 1983. applied geography and new geographies. *Professional Geographer* 35: 88–89.

Moriarty, B. 1978. Making employers aware of the job skills of geographers: a promotional program. *Professional Geographer* 30: 315–318.

Munski, D.C. 1984. Geography academic advisors as foreign language and area studies advocates. *Professional Geographer* 36: 462–463.

Richardson, D.B. 1989. Doing geography: a perspective on geography in the private sector. In *On Becoming a Professional Geographer*, ed. M.S. Kenzer, pp. 66–74. Columbus, OH: Merrill.

Rundstrom, R.A. and Kenzer, M.S. 1989. The decline of fieldwork in human geography. *Professional Geographer* 41: 294–303.

Russell, J.A. 1983. Specialty fields of applied geography. *Professional Geographer* 35: 471–475.

Salter, C.L. 1983. What can I do with geography? *Professional Geographer* 35: 266–273.

Sant, M. 1982. *Applied Geography: Practice, Problems and Prospects*. London and New York: Longman.

Smith, B.W. and Hiltner, J. 1983. Where non-academic geographers are employed. *Professional Geographer* 35: 210–213.

Stutz, F.P. 1980. Applied geographic research for state and local government: problems and prospects. *Professional Geographer* 32: 393–399.

Wilbanks, T.J. and Libbee, M. 1979. Avoiding the demise of geography in the United States. *Professional Geographer* 31: 1–7.

Wynn, G. 1983. Human geography in a changing world. *New Zealand Geographer* 39: 64–69.

Martin S. Kenzer
Department of Geography and Anthropology
Louisiana State University
Baton Rouge, LA 70803–4105
U.S.A.

SECTION I

Ruminations and Philosophical Queries

1. Why Applied Geography?

'Ants are a curious race.'
Robert Frost

Geographers, like Robert Frost's ants, are a curious race. As a group we are woefully insecure, and nothing binds us together more closely than the deep-seated conviction that we are not properly appreciated. So many members of our tribe complain so constantly about the failure of geographers to receive the respect they think we deserve that I once was tempted to wonder whether Herodotus (fifth century B.C.) might have been the first to complain about the poor public image of geography. Certainly he was not the last; Ronald F. Abler, in his presidential address to the Association of American Geographers (AAG) in 1987, felt moved to castigate 'the incessant whining of geographers' (Abler 1987, p. 515).

Their insecurity has prompted a few geographers to propose doing some extraordinarily silly things that they claim will help to improve our image, such as taking out full-page advertisements in national magazines, or hiring skywriters to write something (fuzzy-headedness and specificity are not congenial bedfellows) across the skies at major outdoor sporting events, or buying advertising space on boxes of breakfast cereal, or even hiring a full-time PR person (Hart 1982, 2).

Geographers manifest their profound and pathetic insecurity about their discipline in at least two unfortunate ways. One is the sense of shame that some of them seem to feel when they are identified as geographers. Apparently they would rather be identified as second-rate practitioners of almost any other discipline than as proud members of their own. Geomorphologists try to pretend that they are geologists, historical geographers try to pretend that they are historians, economic geographers try to pretend that they are economists, and a few geographers even try to pretend that they are sociologists.

I become uncomfortable whenever anyone starts inserting adjectives (except 'good') in front of 'geography,' because I see the subject as a seamless web, and it is foolhardy to try to separate the various parts from one another. We all need each other, and no useful purpose is served by erecting artificial barriers within the discipline.

The other unfortunate manifestation of the insecurity of geographers is the contempt that many of them have for their fellows, as evidenced by the two caste systems (or pecking orders) that have developed within the discipline. One caste system is based on specialization, and the other is based on institutional affiliation. Some 'quant men,' for example, have never tried to conceal their scorn for those who are not quantitatively inclined, some physical geographers dismiss all human geography as intellectually soft and undemanding, and those who aspire to be identified as scientists are quite contemptuous of those who do not.

M. S. Kenzer (ed.), Applied Geography: Issues, Questions, and Concerns, 15–22.
© 1989 *Kluwer Academic Publishers.*

Arrogance, scorn, and contempt normally generate an unproductive counter reaction, because most of us have difficulty developing much respect and affection for those who are openly contemptuous of what we are trying to do. Some of the Young Turks of our Quantitative Revolution, for example, made no secret of their disdain for much of the work of traditional geographers, and some traditionalists yielded to the temptation to respond in kind. The attitudes that hardened in both camps created a climate of suspicion and mutual distrust that regrettably still lingers on in some quarters, to the detriment of all geographers.

Far more pernicious is the caste system based on institutional affiliation, which is widely recognized but rarely articulated. At the very top of the heap are the geographers in a handful of major doctoral departments, mostly in large state universities, who look down on everyone else, and just below them are the geographers in other doctoral departments. Somewhere beneath them are the geographers in the regional universities, nee colleges for training teachers, who kowtow properly to their brethren in doctoral departments, but treat all others with the contempt with which they themselves are treated by the geographers in the doctoral departments. Still further down the line are the geographers in liberal arts colleges that do not have graduate programs, and all the way down at the bottom are the Untouchables of geography, those poor souls who do not hold academic appointments of any kind.

Any knowledgeable geographer can immediately think of obvious exceptions to this model of a caste system within the discipline, but, like any other good model, it fits most of the observed phenomena remarkably well. The exceptions generally are the best geographers. They are strong and secure in their status, and they can afford to be generous and gracious to all of their colleagues. Those who cling to the caste system most tenaciously are geographers who are weak, insecure, and aware, however subconsciously, that by some miracle of good fortune they have been elevated to a higher level than they have earned or deserve.

Some years ago, at an AAG meeting, I had a vivid experience of the way in which some academic geographers have treated their nonacademic colleagues like second-class citizens. I ran into one of my good friends, who has his own very successful consulting service. He was almost hopping up and down he was so angry. Seems he had just been talking to the chairman of one of our weaker doctoral departments, and the chairman had said to him, 'I have a graduate student I think you should hire. He is so weak that I can't recommend him for an academic position, but he certainly is good enough to work for you.'

It would have been truly astonishing if such arrogance had failed to generate a vigorous reaction, and nonacademic geographers are only human. They have reacted by apotheosizing the concept of 'applied geography,' which has become a shibboleth, a slogan, a rallying cry, a defense mechanism to protect nonacademic geographers against the shabby ways in which they have been treated by those in Academe who are supposed to be their friends, professional colleagues, and strongest supporters.

Some nonacademic geographers, unfortunately, have been a bit too shrill in making their case, and in turn they have forced academic geographers into defensive postures, which has hurt both parties. There are times when applied geography seems to be in the same category with such matters as abortion, smoking, handgun control, and the proper form of footnotes, which apparently are not conducive to dispassionate and rational discourse. Discussions of applied geography all too often seem to degenerate into emotional outbursts that do more harm than good.

I have implied that applied geography can be equated with nonacademic geography, which is patently absurd, because there is considerable overlap between academic and applied geography. An unbroken continuum stretches from the dustiest recesses of the ivory tower to the wildest pandemonium of the marketplace, and attempting to identify a sharp break anywhere along it would be an exercise in triviality.

No one has ever been able to come up with a definition of applied geography that satisfied anyone but himself or herself, but I would be untrue to my precepts and beliefs if I failed to provide an operational definition of an important term I was using. Therefore, my own working definition – which of course satisfies no one but me, and I reserve the right to change my mind at any time! – says that applied geography is the synthesis of existing geographic knowledge and principles to serve the specific needs of a particular client, usually a business or a government agency.

Almost all geography is applied to some degree, whether in the classroom, in scholarly discourse, in the corridors of government, or in the marketplace, and the possible applications of geography are limited only by the imagination and creativity of the applicator and by his or her ingenuity in finding clients who are willing and able to pay for the services that he or she can render.

Service to a specific client, whether business or public agency, seems to be the common denominator in the thinking of many people when they talk about applied geography, and it also seems to distinguish applied geography from academic geography. Most academics have some sense of the students who are enrolled in their classes, but they rarely think in terms of specific constituencies, much less in terms of specific clients. Most of them do have vague notions about serving society at large, and I suspect that many would agree with Sherman Kent, that 'the scholar who feels his social obligation and who covenants to pay his way in life by research contributes to society's formal store of useful knowledge' (Kent 1946, p. 3).

But should research be useful? At a meeting of historical geographers in Los Angeles in 1979 I got on my soapbox and started to harangue the assembled multitude about the desirability of producing scholarly work that was useful. When finally I ran out of gas, Prof. H.C. Darby, the dean of historical geographers, quietly observed, 'I have never once, in my entire career as an historical geographer, ever done one single thing that was useful.' I chose not to debate the issue, but I thought of the nine-volume Domesday geography, *An Historical Geography of England Before 1800* (Darby 1951), *The Changing Fenland* (Darby 1983), his service on various royal commissions, and I concluded that he and I might have rather different definitions of usefulness.

I must also confess that I preach much better than I practice. I recognize the importance of applied geography to the good health of the discipline, but I have also been remarkably successful in having failed to find any client outside the groves of Academe who was willing to pay for any of the services I might have been able to offer.

I represent the completely academic and impractical tradition in geography. My baccalaureate degree is in the classical languages, Latin and Greek, which did not exactly have job recruiters beating a path to my door, although I do believe that my training in the classical languages has been remarkably helpful in teaching me how to use my own native language more effectively.

I was attracted to geography because it gave me a better understanding and appreciation of the world in which I live, and my greatest pleasure is sharing with others some of the things I have learned. I enthusiastically endorse the view espoused by Darby in his inaugural lecture at the University of Liverpool in 1946 (see Darby 1946): my goal as a teacher of geography is to help students learn to read their morning newspapers with greater intelligence and understanding, and to take their evening walks or their Sunday drives with greater interest, appreciation, and pleasure.

In short, I treasure geography as one of the essential subjects in any liberal education, and for myself I ask no more of it. I believe that helping people to understand and appreciate the world in which we live is an important application of geography. People need geography, but geographers cannot blithely assume that they are aware of this need. We have a responsibility for educating them to it. One of my major tasks, as a teacher of geography, is to demonstrate its value and applicability to the daily lives of those I am trying to teach.

Those of us who teach college geography must learn to do an even better job of selling our subject and ourselves, both to students and to the body politic, no matter how distasteful we may find such an assignment. We need students, and we need majors. Perhaps a few institutions of higher education in this country will be the lucky exceptions, but the great majority will continue to be enrollment-driven, and recruiting students into our classrooms will remain essential to our survival as an academic discipline. We must develop a loyal and supportive constituency.

Geographers may have an advantage, paradoxically, because we have not enjoyed a place of privilege in the curriculum. We have not been able to rely on hordes of 'captive' students who are required to take our courses, and we have accordingly been forced to attract students by the superior quality of our teaching. It is dispiriting to have to teach large masses of disinterested students who have enrolled in a course because someone ordered them to do so, and it is my impression that the quality of instruction in large required courses is generally inferior.

I also have the impression that most geography departments, perhaps out of necessity, do a better job of teaching than most of the other departments in their institutions. I am sorry to have to add that this may be a commentary on the pathetic quality of instruction in those other departments, because there is considerable room for the improvement of instruction in geography. Being the best of a sorry lot is nothing to write home about.

Geographers can also recruit good major students by holding out the prospect of nonacademic careers that are fulfilling. I personally happen to dislike the current emphasis on vocationalism in American higher education, but I seriously doubt that my dislike is going to change it, and geography is more fortunate than many other disciplines in the social sciences and humanities in this regard, because we do teach students techniques that give them easy entry into the job market.

Once upon a time college teaching was the principal career opportunity for those with liberal arts degrees. Colleges of liberal arts all pay lip service to the ideal of educating the citizenry, but in fact they have traditionally been in the business of reproducing themselves by training prospective college teachers. Their graduate programs, in particular, have been as professionally oriented as those of any other units in the university. The profession for which they have been training their students has been college teaching, and they suddenly found themselves up the creek in the early 1970s when the bottom dropped out of the academic job market because the last of the baby boomers had graduated and undergraduate enrollments were starting to decline.

Academics had to learn to live with the fact that the traditional market for teachers was not as great as it once had been, and it probably will never return to its former glory. We had to adjust to the changing times by encouraging our students to pursue alternative, nonacademic careers (Hart 1972, p. 14), but many academic geographers, and I am one, are woefully ignorant of the world outside our ivory tower. We did not know where or to whom to turn for advice in adapting our curricula to fit the new circumstances, and a shocking number did not even seem to care. They elected to play ostrich, buried their heads in the sand, bared their derrieres to the world, and tried to pretend that nothing had changed.

The more thoughtful scholars recognized the necessity of rethinking their role in society and of paying greater attention to the needs of clients outside Academe. Our teaching had to be tailored to serve these needs more adequately, and we had to become more conscious of our own responsibility for serving them directly through our research and our service activities.

Service, teaching, and research are the hallowed triad to which all university administrators pay lip service, and often little more. Many faculty members misconstrue service to mean no more than doing time on university committees, although such 'service' is merely enlightened self-interest, because most academic committees serve little useful purpose other than the socialization of the faculty. Junior faculty members should feel obligated to serve on them as a way of becoming acquainted with colleagues in other departments, and senior faculty members should serve on them to get acquainted with junior colleagues.

The service mission of a university entails service to the community, to society, and to the state, the region, and the nation. We should pay full homage to the handful of geographers who have compiled creditable records of public service, but very few academic geographers are psychologically attuned to its demands. It is fearfully time-consuming, and we begrudge the loss of time from our teaching and research. We are not prepared to make quick decisions, even when decisions must be made quickly, until we have marshalled all of the necessary evidence; but once we have reached our conclusions we are quite impatient with the interminable process of review, compromise, and revision.

If universities and the body politic truly believe that service is important, and I
think that most of us would agree that it is, it seems to me that all scholars, and not
just geographers, have an obligation to make their talents and some of their time
available to serve the needs of society. All geographers, in short, should do their
best to become applied geographers, at least part-time, but each of us in accordance
with our own individual lights.

Applied geography, however one chooses to define it, is essential to the good
health of the discipline, but we would be foolish to ignore the fact that, from a
strictly scholarly perspective, it suffers from two serious weaknesses: advocacy and
parasitism. First, it is difficult for an applied geographer to avoid becoming an
advocate for a particular position, and second, applied geographers may be
described as intellectual parasites who draw from the larder of the discipline but do
little to enrich that larder.

Much applied geography attempts to influence public policy, often on behalf of a
particular client, and the applied geographer may be tempted, perhaps even
unconsciously, to put the interests of the client before the interests of society. There
is a very real danger that applied geography can lapse into advocacy. Scholars lose
their credibility when they become advocates, because the scholar is expected to
weigh all evidence as objectively as possible, but the advocate is expected to make
the strongest possible case for a client, and the advocate minimizes, ignores, or
even suppresses evidence to the contrary. You know what advocates are going to
say before they even open their mouths, and you automatically suspect their
conclusions.

The other serious weakness of applied geography is its intellectual parasitism.
Applied geographers depend on the knowledge, ideas, skills, and principles that
have been developed by other geographers, but normally they do not share the
results of their own work with their colleagues, and the rest of us simply cannot be
privy to what they are doing. They do not contribute to the organized assault on the
frontiers of knowledge, despite the fact that any discipline will eventually wither
away and die if it is not infused periodically with new ideas.

I can identify three reasons for the failure of applied geographers to enrich the
literature of the discipline. The first is security. Many applied geographers routinely
use classified, proprietary, or otherwise privileged information, and they are not
free to disseminate the results of their work even if they wish to do so.

The second reason is the lack of any pressure or reward for scholarly publication.
Apparently some people have to be rewarded before they will do anything. I find
this argument singularly unconvincing, because the good scholars I know feel
compelled to publish because they are so eager to share what they have learned,
and the rewards of publication are merely incidental.

I think the most important reason for the failure of nonacademic geographers to
contribute to our scholarly literature is what I will call, for lack of a better term, the
intimidation factor. Many nonacademic geographers are simply afraid to submit the
results of their work to our scholarly journals because their paper-writing skills
have grown rusty through disuse since they left graduate school.

Now, I realize that applied geographers regularly prepare reports for their clients, and those in public service are quite familiar with the experience of seeing their handiwork tortured beyond recognition by layer upon layer of bureaucratic review within their agencies, but they have little practice at or experience of the skills required to prepare their ideas for scholarly publication, or of the often painful indignities of seeing a paper into print in a scholarly journal, although the review process in a public agency can be crueler, harsher, and even more witless than the review process for scholarly publication.

Applied geographers are not alone in their failure to enrich the literature of our discipline, because teachers in the elementary and secondary schools seem to suffer from the same intimidation factor. The *Journal of Geography* is supposed to be written by and for teachers, but every editor of the *Journal* that I have known has complained about his inability to elicit an acceptable flow of manuscripts from teachers.

I am saddened that the intimidation factor deprives the rest of us of many of the good ideas of applied geographers and of elementary and secondary school teachers. I wish the members of both groups would contribute more to our scholarly literature, because I have been greatly enriched by my own conversations with them, and I fear they may be classic manifestations of the much abused and grossly misunderstood cliche, 'publish or perish.' It really means that your ideas will perish with you unless you have put them into print.

In summary, I have asserted that all geography is applied to some extent, whether in the classroom, in scholarly discourse, in the corridors of government, and/or in the marketplace. In a narrower sense, applied geography abstracts knowledge and principles from the corpus of geographical scholarship and synthesizes this knowledge and these principles to satisfy the specific requirements of a particular client, usually a business or a government agency. The client often forbids dissemination of the results of the work that has been done for it, and thus this work fails to enrich the literature of the discipline. Other geographers perforce remain ignorant of much of it, which is most unfortunate.

It is equally unfortunate that the concept of applied geography has been divisive and harmful to a very small profession. Some nonacademic geographers have felt compelled to invoke applied geography as a defense mechanism to protect themselves against the arrogance and scorn with which they have been treated by some academic geographers, and the flourishing of applied geography has generated a counter reaction from some academic geographers.

A discipline as diverse as geography can and should make use of many different kinds of inclinations and talents. We should be permissive, not exclusive, and we do not need more rigorous tests for orthodoxy. Geographers should lay aside their petty squabbles, and each one of us should make a sincere and earnest effort to develop a better understanding, more tolerance, and greater mutual appreciation and support for the work of all of the other members of our clan.

Acknowledgment

The author is grateful to Liz Barosko for the skill with which she typed the manuscript.

References

Abler, R.F. 1987. What shall we say? To whom shall we speak? *Annals, Association of American Geographers* 77:511–524.

Darby, H.C. 1946. *The Theory and Practice of Geography*. London: University Press of Liverpool; Hodder & Stoughton.

Darby, H.C., editor. 1951. *An Historical Geography of England Before A.D. 1800.* Cambridge: Cambridge University Press.

Darby, H.C. 1983. *The Changing Fenland.* Cambridge: Cambridge University Press.

Hart, J.F. 1972. *Manpower in Geography: An Updated Report.* Commission on College Geography Publication No. 11. Washington, DC: Association of American Geographers.

Hart, J.F. 1982. The highest form of the geographers's art. *Annals, Association of American Geographers* 72:1–29.

Kent, S. 1946. *Writing History.* New York: F.S. Crofts and Co.

John Fraser Hart
Department of Geography
University of Minnesota
Minneapolis, MN 55455
U.S.A.

2. Homilies for Applied Geographers

Introduction

The idea of applied geography seems especially alluring, both theoretically and practically. Theoretically, it seems to lend an immediacy and robustness to geographical study. As Hare put it:

> Geography for its own sake does little for my personal satisfaction. Geography conceived as a guide to sensible human action does a great deal (1977, p. 262).

Practically, promises of public money, prestige, and security exert a powerful influence, especially if it appears that applied geography can be practiced from the socially protected sanctuary of the university.

Yet, geographers seem to have awakened relatively late to the idea of applying their disciplinary knowledge to public problems. During the 1950s and 1960s, despite the prodigious efforts of Stamp (1960), geographers turned inward, away from such problems and geography became more theoretical, more self-absorbed, more methodologically diverse, something descriptively, if sarcastically, called 'cocooning' by Soja (1987). A forum devoted to applied geography did not appear until 1981 with the journal *Applied Geography*. By contrast, during the same period many economists, political scientists, sociologists, and decision theorists were already boldly proclaiming the usefulness of their disciplinary knowledge in solving public problems – a series of claims that Lasswell summarized under the head 'policy science' (1951). Lasswell and his followers thus attempted to put a disciplinary stamp on involvement of social scientists in government, a development that began during the late nineteenth century, accelerated during the New Deal and World War II, and has since become a burgeoning industry (see Silva and Slaughter 1984).

Perhaps it would be more accurate to say here not that geographers were *awakened* to the idea of application, but rather that, with the passing of generations and fading of early enthusiasms, they were *emboldened* to pursue it anew. Government bureaus, research institutes, journals and other publishing outlets devoted to policy science have proliferated.

Fatalists claim that, sooner or later, all scientific knowledge is applied. To them, application of geography would seem somehow inevitable. But, this is too easy. It gives an historical answer to what really are political and moral questions; and, in doing so, it invites hubris, intellectual dishonesty, and moral confusion. These are sins we all want to avoid; but they are easy to commit. Conceptual, moral, and political traps abound in efforts to apply disciplinary knowledge. What follows, then, is a modest attempt to assist in avoiding such traps, if only by pointing out some ways to fall into them. It is constructed in the form of three homilies. These

M. S. Kenzer (ed.), Applied Geography: Issues, Questions, and Concerns, 23–34.
© 1989 *Kluwer Academic Publishers.*

homilies are not, however, mere moralizing lectures but rather methodological and philosophical cautions, some drawn from witnessing the development of disciplines that embraced the application idea early on. We call them:

1) thin conceptualization,
2) deflationary ethics, and
3) intellectual self-immolation.

Homily 1: Thin Conceptualization

The editorial in the initial issue of *Applied Geography*, which sets out the 'principles and practice of applied geography,' is a good place to begin. The editorial talks about 'scarce resources,' about the exigencies of 'human existence,' about exploiting such resources to 'best long-term advantage,' about finding 'solutions.' Indeed, the passage from Hare quoted at the first of this paper, which extols 'sensible human action,' heads the editorial (1981, p. 1).

Unfortunately, none of these concepts – scarce resources, human existence, best long-term advantage, sensible human action – is politically or ethically neutral. They are so thinly conceived that they may appear to be, but that is misleading. Talk of exploitation of resources to the best long-term interests of humanity, for instance, is altogether too abstract and grand; it adumbrates a homogeneous conception of human interests that is inappropriate to the facts of the actual world in which we live (see below). The entire earth and its inhabitants are conceived as some ideal community (though one, perversely, evidently not recognizing itself as such). 'Sensible human action' and 'solution' are similarly misleading. What is sensible for some is folly for others; what is a solution for some is a problem for others.

Parallels with the ways in which policy analysis and planning have conceived their tasks will help bring out the actual content of such thin conceptualizations in policy contexts. In effect, applied geographers *become* policy analysts and planners.

Policy analysts and planners also have formulated the objectives of their enterprises in a substantively empty way, as contentless systems of analytical techniques. As Lasswell, who introduced the term *policy science* into American political discussion, put it, policy science is 'knowledge *of* the policy process and of the relevance of knowledge *in* the process.' It embodies a distinctive outlook: contextuality, problem orientation, and technique synthesis (Lasswell 1971, p. 3). Similarly, in his review of policy study, Nagel claims that the main mission of policy analysis 'is to help in identification of preferable solutions to present pressing policy problems' (1973, p. 2). And, in the introductory editorial to the new journal *Policy Sciences*, Quade states that:

> The intention of the policy sciences is simply to augment, by scientific decision methods and the behavioral sciences, the process that humans use in making judgments and taking decisions (1970, p. 1)[1]

Planners developed a similar view of their task. Thus Davidoff and Reiner described planning as:

...a process for determining appropriate future action through a sequence of choices ... since appropriate implies a criterion of making judgments about preferred states, it follows that planning incorporates a notion of goals. Action embodies specifics, and so we face the question of relating general ends and particular means.

The choices which constitute the planning process are made at three levels: first, the selection of ends and criteria; second, the identification of a set of alternatives consistent with these general prescriptions, and the selection of a desire alternative; and third, guidance of action toward determined ends (1962, p. 103).

Here, planning, like policy analysis, is considered a substantively empty system of analytical techniques; also, like policy analysis, it is conceived as oriented to problems, but in an abstract and completely indeterminate way. Although such general categories as public interest and social welfare are included, they are not made specific or problematized within policy analysis and planning themselves.

In this way, then, the American founders' abiding worry about a Machiavellian Moment (Pocock 1975), in which republics were thought inevitably to confront their own mortality, has been wholly replaced by policy analysts' and planners' assurances of a Lasswellian Moment – i.e., the enduring present, in which there is always basic agreement on broad social and political objectives, which can be exhaustively cataloged by proper analysis, and continual progress in achieving them through application of policy-analytical technique.

Another way to see this is to read catalogs of public affairs schools (most of which were established in the late 1960s and 1970s). Older, more traditional disciplines, such as economics, political science, geography, and anthropology may seem to be exempt from the charge of innocence, but this will not withstand scrutiny. Diesing (1982), for instance, uses the term 'policy sciences' in a sense restricted precisely to these disciplines. Although this is not by itself persuasive, it is clear that the line between them and the more practice-oriented approaches is very thin. In fact, the claims to knowledge in the latter rest on the knowledge claims of the former. It was just in this spirit that Lasswell predicted, somewhat naively, that:

...the policy-science orientation in the United States will be directed toward providing the knowledge needed to improve the practice of democracy (1951, p. 3).

This, despite the fact that the meaning of *democracy* has been a major center of political controversy and conflict throughout American history.

Such innocent formulations are vitiated from the start by abstraction away from actual systems of human practices in actual historical contexts, and by failure to address the concrete social, economic, and political forces that have called the institutions of social control geographers wish to join into existence in the first place. Obviously, then, these formulations are anything but neutral. They contain a political view and a very conservative one at that. They assume that basic political questions about how best to organize society have already been answered (in major outline anyway), and furthermore assume that consensus exists about that answer, so that all that remains are technical difficulties in making that sort of organization work. The roles of policy analysts and planners are simply assumed to be those of systems-modernization and system-maintenance agents.

Such formulations avoid questions raised, for instance, by radical geographers about the role of capitalism in shaping spatial organization and 'continuously revolutionizing the geography of production, exchange, and consumption, and so closing off alternative paths of development' (Harvey 1985, p. 83). More generally, they avoid questions raised by environmental, peace, urban, feminist and other social movements, which express an increasing awareness of the blind, self-destructive effects of modernization and rationalization. They are not Luddite, but on the contrary, project the view that modernization processes relying exclusively on instrumental and strategic rationality are inherently counterfinal and destructive: greater military security increases likelihood of war; increased production and consumption destroy the physical environmental effects; unchecked urban growth destroys neighborhoods; more and more areas of social life come under surveillance; political participation becomes merely formal. These movements thus project a *critique* of modernization and system conservation efforts (see Habermas 1981a, 1987; Offe 1985).

Plainly, therefore, these thin conceptualizations really suggest a political ideology. It is a strikingly anti-democratic ideology.[2] Contrary to claims of Lasswell and other policy scientists, it is profoundly hostile to democratic principles of popular consent, equality, and participation. It does in fact make substantive value assumptions about American democracy and about democratic political process. It is elitist and exclusionary. It lacks any semblance of democratic legitimacy. And, it can actually encourage political manipulation through outright deception or more often through what Goodin calls 'manipulation of the obvious,' through linguistic debasement, problem delineation, fact-fiddling, and concept-rigging (Goodin 1980, p. 196). In sum, thin conceptualization deprives applied geographers, as it does policy analysts and planners, of any independent critical perspective from which to understand their own enterprise and, by extension, to understand themselves.

It may seem tempting here to fall back on the claim that science is neutral in such respects. This seems much easier to claim in physical science than in social science, though it is becoming harder there too. It is very difficult indeed when dealing with concrete policy problems. Very often there is the awkward difficulty that choice of one policy over another will have very different distributional and other consequences. Under one choice, some people will benefit and some will be harmed; under another, different people will benefit and different people will be harmed. Sometimes such choices can even affect the number of people who will later live (see Parfit 1984).

Lasswell explicitly disclaims neutrality. What this means in context, however, is that policy analysts are not to intrude their values into policy discussion but instead are to take on the values of their clients. Thus, values become just another analytic variable in analysts' equations; the neutrality question is not so much solved as hidden against the background. On this view, policy analysts are virtual political and moral chameleons who take on the protective coloration of the policy environment in which they find themselves.

In any case, all claims to political neutrality finally wreck on what might be called the too-many-chiefs-and-not-enough-Indians mistake. Political neutrality in fact embodies a fallacy of composition: the fallacy that what is possible for any single individual must be possible for them all simultaneously. But, not everyone can be politically neutral. Here the elitism of special claims for expertise is clearly exposed.[3]

Homily 2: Deflationary Ethics

Evaluation of public policy programs usually invokes notions of the public interest and public welfare. These are generally given a utilitarian twist: utility becomes the central evaluative concept. There is some logic to this. Public projects and programs typically affect broad groups of people. They are largely concerned with outcomes and, accordingly, a consequentialist ethics centered on utility appears appropriate. Nevertheless, this has its dangers too. It is a great oversimplification and easily becomes reductionist and deflationary, particularly with respect to the strategy (especially prominent in choice theory) of searching for some social welfare function that will aggregate the preferences of all affected by a policy decision.

From a policy perspective, the difficulties, apart from the logical problems internal to the notion of a social welfare function pointed out by Arrow (1963), lie in the fragmentation of value and the complexity of interests. Value has many roots, not just one; human interests are multidimensional, not unitary.

Many, rather than one, value standpoints seem possible: economic, political, and personal liberty; equality; equity; privacy; procedural fairness; intellectual and aesthetic development; community welfare; desert; non-arbitrariness; risk acceptance (or aversion); welfare of future generations; interests of other countries; and so on. Furthermore, these appear heterogeneous. Some are more person-centered, others more outcome-centered; some are more objective, others more subjective.[4] They might be grouped into the following rough categories:

Obligations	person-centered	objective
Rights	person-centered	objective
Utility	outcome-centered	objective
Perfectionist Ends	outcome-centered	subjective
Private Commitments	person-centered	subjective

Whether this classification is exhaustive or not, the point is that utility is plainly not the only moral consideration. People can, and do, morally represent the world in many different ways at the same time.

Fragmentation of value reflects the multidimensionality of human interests. People have a number of specific interests, not just a general interest in doing as well as possible, all things considered. These interests include being allowed to speak freely, being able to own property, not being objects of racial discrimination, not being imprisoned, and so on. These interests are not, however, aggregatable into some more fundamental interest. The space of peoples' interests has multiple dimensions and lacks a natural metric. Thus, people do not always regard an insult to one interest as compensated for by a gain in some other interest. Many moral requirements are concerned simply to protect certain interests and are not concerned with how an action affects a person's interests as a whole. Hence, to show that an action is wrong it will be enough to show that it adversely affects some one specific interest protected by a right. One consequence is that peoples' preferences for various possible actions of others that will affect their interests will not yield simple, or complete, orderings but a much more complex collection of incomplete partial orderings (see Woodward 1982).

Reductionist approaches thus distort critical features of human choice situations and encourage systematic mistakes in moral mathematics (see Parfit 1984).[5] In many cases, perhaps most, it actually is impossible to bring different sets of consideration for possible actions together in a single decision. Yet decision is, of necessity, single. Reductionist approaches actually *falsify* situations where there are genuine conflicts in choice. In many policy situations, perhaps most, multiple considerations will be relevant. Each may have different justificatory roots; obviously not all will be compatible nor can be made to seem compatible. (This can be said, *mutatis mutandis*, for nonmoral considerations as well.) Utilitarian calculi – cost-benefit analysis, and its generalized version, risk-benefit analysis, futures research – institutionalize this distortion. These techniques in fact embody a double abstraction from actual choice situations: first, abstraction from their essential complexity and messiness by imposing a single utility calculus; and, second, through monetization, by imposing an external and homogenous measure of value and reducing diversity to a single comparable dimension. Objective human social and political relations are thereby masked by objective economic relations.

These reflections have serious implications for anyone involved in public policy. Specialization of function and narrowing of institutional roles worsen chances that such mistakes in moral mathematics will be noticed and somehow accounted for in public decisions. Yet, with decision comes accountability, so that those involved in decisions are confronted with a modern version of Machiavelli's problem of dirty hands.[6] Here it is useful to remember that the great crimes of this century have been public crimes and that expert advisors were either directly involved in their perpetration or worked in the shadows of the perpetrators.

Perhaps public policy is best understood as a search for reasonable bases of agreement among people with sometimes very different interests and preferences, bases that can be impartially defended and agreed to by all. The communicative ethics of Habermas would thus provide a better model for public policy than utilitarian calculi of social choice theory (Habermas 1973, 1976, 1981b).

A related question is whether people have real interests that differ from the ones they think they have. Public policy practitioners want to say yes. They want, for instance, to distinguish between *felt needs* and *professionally recognized needs* (see, e.g., Kueckeberg and Silvers 1974, p. 9). Whether this is a sound distinction is controversial (see Davidson 1985 for an argument against it). Perversities in belief and preference formation, including hysteresis, suggest that some such distinction is necessary, both theoretically and practically. Nevertheless, in light of the value fragmentation and interest prolixity discussed above it is not at all clear what follows from it for public policy. Certainly the wholesale interventions into peoples' lives advocated, for instance, by Goodin (1982, pp. 39–58) are not justified by it. The mere fact of a distinction is not a justification; it is itself a political and moral problem. It is certainly a model that takes democracy seriously.

Homily 3: Intellectual Self-Immolation

A major pitfall in linking disciplinary knowledge and public policy is loss of scientific autonomy, a sort of intellectual self-immolation. All too easily, funding becomes the major factor in choice of problems and methods; claims of scientific truth are eclipsed by non-scientific demands of the larger social and economic system.

A compelling, if mythical, view of science is captured in the Popper-Kuhn-Lakatos-Laudan accounts of scientific change: that science is autonomous. The autonomy view is clearly stated by Polanyi, who claims that the body of scientists as a whole constitutes a 'Society of Explorers.' This he calls the 'Republic of Science':

Such a society strives towards an unknown future, which it believes to be accessible and worth achieving. In the case of scientists, the explorers strive towards a hidden reality, for the sake of intellectual satisfaction. And as they satisfy themselves, they enlighten all men and are thus helping society to fulfill its obligation towards intellectual self-improvement (Polanyi 1962, p. 72).

But this seems naive. What we actually observe is much different. In fact, modern natural and social science are just other sets of institutions among many others in a larger, highly differentiated society. They are professionalized and subject to all kinds of social, political, and economic pressures.

Toulmin paints a more accurate picture. He sees science as a *tertiary industry*. When the economic history of Western society over the last 150 years is looked at from the perspective of occupation and employment, the following picture emerges:

A century and a half ago, the vast mass of the work force in all countries was employed in agriculture, forestry, mining, and fishing – *primary* industries. Only during the last 50 years has the proportion of the work force so engaged in the major industrialized countries dropped sharply, from well over 1:2 to something less than 1:10. To begin with, though with periodic lapses, the laborers who were not needed for *primary* production found occupation in manufacturing industry and in the associated clerical and service trades. They did not them-selves produce new materials directly but they did process them into saleable objects – or else they engaged in other activities (e.g., bookkeeping, stock-taking, trucking, cleaning, maintenance) ancillary to such processing. So there came about an age of *secondary* industries.

But this has proven to be only one more phase in economic history. As Keynes foresaw, the increasing efficiency of manufacture presented a choice: either employment in producing unwanted goods or employment in some other range of activities, perhaps more creative ones. Toulmin continues:

Pure scientific research is, and can deliberately be chosen to be, one of those new, *tertiary* activities by which employment and prosperity can be maintained in an industrialized society, even after both primary and secondary industries have become *too efficient* to occupy the available labor-force. A scientific research laboratory [or a university], just as well as a manufacturing enterprise, is the focus around which the life and prosperity of a community can be organized (Toulmin 1966, pp. 163–64).

Toulmin puts a somewhat hopeful cast on this. If his picture is accurate, however, in today's post-industrial world science is simply an instrument for aggregating resources and developing solutions employed by complexes of private corporations and public institutions: an instrument that can be used in any area, for any purpose, for any need, and against any contingency.[7] The basic method for allocating research in this system is the policy project. Such projects attempt to appropriate future resources by identifying social problems in such a way that the expertise of a particular complex is believed necessary to its solution. Manifestly, under such arrangements science is committed neither to discovery, to disciplinary consensus, nor to service to society (Blisset 1972, p. 162). Science is, in short, manipulated by the guidance system of the post-industrial state that Galbraith calls the *technostructure*.

In a technologically driven society, human and physical resources increasingly become the object of state policies of modernization and rationalization, and are increasingly defined as falling under policy scientific expertise. Clearly, scientists are drawn into public decision-making partly because of the complexity of the issues and consequent dependence of decision makers on scientific expertise; still, their expertise can easily be used to serve the political functions of *rejecting* claims, or potential claims, of non-experts to be heard:[8]

...as soon as an issue is institutionally defined as requiring *scientific* advice and expertise, the scope of legitimate participants is drastically reduced ... By replacing democratic procedures of consensus building by such other methods of conflict resolution, government elites *avoid* the 'official' institutions of politics in a constant search for *non-political forms of decision-making* (Offe 1984, p. 168).

As political decision-making shifts from the political forum to the boardroom and office, the scope for consensus-building politics narrows. Not only do such corporatist methods of decision-making have no clear democratic legitimacy, they are in fact hostile to democracy. People learn to participate in consensus-building politics partly by participating. It is through such participation that the sense of political efficacy essential to democracy is formed and nurtured. So their exclusion by claims to expertise has a double effect: it actually excludes them and also undermines their abilities to participate. They become more politically dependent and more and more manipulated. Offe offers a metaphor that graphically captures this point:

> ...compare a chicken living in the natural environment of a farm to a chicken being raised in the technologically advanced environment of a modern chicken factory. It is clear that the latter, deprived of the opportunity to practice its instincts which lead it to control and adapt to its physical environment, becomes dependent upon all kinds of support systems supplying it with the right food, temperature, amount of fresh air, infrared light, antibiotics, etc. It would be absurd here to speak here of any increased needs as constituting 'rising expectations' or increased demands, whereas it is obvious that they result from utter helplessness and dependency (Offe 1984, p. 165).

Concluding Remarks

To reprise these homilies:

Homily 1. Thin conceptualizations of the nature of applied geography mask out its actual political nature and encourage belief that applied geographers are detached, neutral observers, mere advisors to political decision makers and not actual participants in the politics of a larger society. This not only fosters self-deception but also, because of the ideological power of science, it can distort communication of scientific results with others, politicians and citizens alike (see Feyerabend 1978; Heilbroner 1985).

Application of disciplinary knowledge *is* political. There is no neutral ground. Instrumental, means-oriented research, moreover, inevitably becomes conservative and systems-maintaining. And it deprives those involved in public policy of any independent critical perspective from which to discuss larger questions about ends.

Homily 2. Complexity of human interests and fragmentation of value make the political world extremely opaque and messy. It is highly resistant to the sorts of utilitarian, consequentialist calculi (such as risk-benefit analysis and futures research) commonly deployed in public policy. In this light, paternalistic intervention is highly problematic morally. Policy advisors, including applied geographers, who participate in such intervention, directly or indirectly, therefore cannot escape moral accountability, no matter how remote the effects may seem. In this, as in most other areas today, it is not just what we do individually but what we do together that matters.[9]

Homily 3. Changes in relationships between science and government threaten the intellectual autonomy of scientists and distorts scientific choice itself. Science can easily become an instrument used by complexes of private corporations and public institutions for appropriating future resources and closing off alternative choices. The effect of such institutional changes is to make questions about application of disciplinary knowledge even more problematic, intellectually, morally, and politically.

In today's technologically sophisticated, crowded, and interdependent world, the question then is not whether to apply disciplinary knowledge but how to do it in ways that are not self-deceptive, politically manipulative, or morally repugnant – ways that do not increasingly turn us all into factory chickens. Finding such ways must, therefore, be as much a part of applied geography as refinement of disciplinary knowledge.

Notes

1. An interesting fact is that most policy studies journals were launched in the late 1960s and early 1970s.
2. In outline, this ideology appears to contain at least the following elements:
   1) Detachment and value-neutrality;
   2) Sanctification of present American government as the apotheosis of democracy;
   3) Instrumentalism;
   4) Existence and efficacy of a social scientific technology;
   5) Superiority of expert knowledge to democratic political process; and
   6) Guided instead of open communication (see Feyerabend 1978).
   The most important points made in calling a system of belief 'ideological' lie in its content and in the willingness of those who hold it to bring it to the surface – to spell it out, in effect – to reflect on it, and to criticize it. Fingarette's discussion of self-deception is helpful in making this latter point clearer. Rather than take for granted explicit consciousness of one's engagement in the world – what one takes as a human subject; how it is one takes or finds one's world, including one's self; the projects taken on; the way the world presents itself to be seen, heard, felt, enjoyed, feared, manipulated, or otherwise experienced – we should take its absence for granted. Ability to spell-out this engagement in the world thus appears as something learned rather than natural. Moreover, it appears as something for which there must be a special reason for spelling-out an engagement. Logically, these reasons must be cast in terms of an individual's aims, motives, attitudes, and understanding of the world and his or her identity. To become conscious, therefore, in this strong sense of 'conscious,' is actually to be engaged in spelling-out that which one is conscious of. Inability to spell-out, which is characteristic of self-deception, is not necessarily inability, or even lack of skill, but rather 'adherence to a policy (tacitly) adopted' of *not* spelling out. It is, as Fingarette puts it, 'a self-covering policy' (1969, pp. 40–42, *passim*).
3. For an interesting discussion of the fallacy of composition, see Elster 1978.
4. Here I follow Nagel (1979). Some might want to add to this list. Thus, Fisk adds class interest (1980). Throughout, I understand person-centered and agent-centered to be equivalent.
5. The general dangers of reductionist approaches should be obvious. On the one hand they may encourage a sort of defeatism that rejects rational theory altogether; on the other hand they may encourage rejection of considerations that cannot be made to fit within the reductionist scheme.
6. For a useful discussion, see Walzer 1976. The term 'dirty hands' actually comes from Sartre.
7. Blisset characterizes *complexes* as 'loosely coordinated centers of decision-making, involving both public and private components, that engage in activities of national scope' (1972, p. 193). He cites pesticide policy as an illustration, but many areas of national policy today fit this characterization.
8. The social and policy science literature in this area is large. See Pigou 1932; Mishan 1979; Marglin 1963; Leonard and Zeckhauser 1985.
9. Here is an imaginary example, modernized from a peasant-bandit example by Glover

(1975). Twenty hungry workers are having their lunch at a construction site. Each has a box of twenty chicken nuggets. A band of twenty hungry urban guerillas happen by and, at knifepoint, each guerilla takes one worker's box and eats all the nuggets. Each worker is left with an empty box and no lunch. Each has been perceptibly harmed. Weeks later, the guerillas, again hungry, notice the same workers having another chicken-nugget lunch. After the first incident, one guerilla, having read a book on ethics, complained of the great harm each had done and proposed that, since eating a single nugget would do no perceptible harm to any worker, this time each guerilla takes only one nugget from each worker's box. Each then did so and all left, each pleased with herself that she had done no harm to any worker. But, again, this is very bad moral mathematics. Together the guerillas did as much harm as each had done the first time.

References

Arrow, K. 1963. *Social Choice and Invididual Values*. 2nd edn. New Haven: Yale University Press.
Blisset, M. 1972. *Politics in Science*. Boston: Little, Brown and Company.
Briggs, D.J. 1981. Editorial: The principles and practice of applied geography. *Applied Geography* 1, 1–8.
Davidoff, P., and Reiner, T.A. 1962. A choice theory of planning. *Journal of the American Institute of Planners* 28: 103–115.
Deskins, D.R. *et al.* 1977. *Geographical Humanism, Analysis, and Social Action*. Ann Arbor: Department of Geography, University of Michigan Press.
Diesing, P. 1982. *Science and Ideology in the Policy Sciences*. New York: Aldine.
Dror, Y. 1973. *General Policy Science*. In Nagel 1973.
Elster, J. 1978. *Logic and Society: Contradictions and Possible Worlds*. Chichester: John Wiley and Sons.
Feyerabend, P. 1978. *Science in a Free Society*. London: Verso.
Fingarette, H. 1969. *Self-Deception*. London: Routledge and Kegan Paul.
Fisk, M. 1980. *Ethics and Society*. Brighton, Sussex: Harvester Press.
Galbraith, J.K. 1968. *The New Industrial State*. New York: New American Library.
Glover, J. 1975. It makes no difference whether or not I do it. *Proceedings of the Aristotelian Society*, Suppl. Vol. 49.
Glover, J. 1977. *Causing Death and Saving Lives*. Harmondsworth, Middlesex: Penguin Books.
Goodin, R. 1982. *Political Theory and Public Policy*. Chicago: University of Chicago Press.
Habermas, J. 1973. *Legitimation Crisis*. Thomas McCarthy, trans. Boston: Beacon.
Habermas, J. 1976. What is universal pragmatics? In *Communication and the Evolution of Society*. Thomas McCarthy, trans. Boston: Beacon.
Habermas, J. 1981a. New social movements. *Telos* 49: 33–37.
Habermas, J. 1981b. *Theory of Communicative Action. Volume 1: Reason and the Rationalization of Society*. Thomas McCarthy, trans. Boston: Beacon.
Habermas, J. 1987. *The Theory of Communicative Action, Volume 2: Lifeworld and System: A Critique of Functionalist Reason*. Thomas McCarthy, trans. Boston: Beacon Press.
Hanson, R. 1985. *The Democratic Imagination in America*. Princeton: Princeton University Press.
Hare, F.K. 1977. Man's world and geographers: a secular sermon. In Deskins *et al.* 1977.
Harvey, D. 1975. *Social Justice and the City*. Baltimore: Johns Hopkins University Press.
Harvey, D. 1982. *The Limits to Capital*. Oxford: Basil Blackwell.
Harvey, D. 1985. *The Urbanization of Capital*. Oxford: Basil Blackwell.
Haynes, K.E., and Stubbings, R.G. 1985. Rationality and relativism: an ecological account. *Scientific Geography Newsletter*. June, 1985.

Haynes, K.E., and Stubbings, R.G. 1987. Planning and philosophy. *Journal of Planning, Education and Research* 6: 75–87.

Heilbroner, R.L. 1985. *The Nature and Logic of Capitalism*. New York: Norton.

Krueckeberg, D.A., and Silvers, A.L. 1974. *Urban Planning Analysis: Methods and Models*. New York: John Wiley and Sons.

Lasswell, H.D. 1951. The policy orientation. In Lerner and Lasswell 1951.

Lasswell, H.D. 1971. *A Pre-View of the Policy Sciences*. New York: American Elsevier.

Leonard, H.B., and Zeckhauser, R.J. 1985. Cost-benefit analysis and the management of risk: philosophy and legitimacy. In McLean 1985.

Lerner, D., and Lasswell, H.D. 1951, eds. *The Policy Sciences: Recent Developments in Scope and Method*. Stanford: Stanford University Press.

Marglin, S. 1963. The social rate of discount and the original rate of investment. *Quarterly Journal of Economics* 77: 95–111.

McLean, D., ed. 1985. *Values at Risk*. Totowa, NJ: Rowman and Allenheld.

Mishan, E.J. 1979. Evaluation and life and limb: a theoretical approach. *Journal of Political Economy* 79: 687–705.

Nagel, S.S., ed. 1973. *Policy Studies in America and Elsewhere*. Lexington, Massachusetts: Lexington Books.

Nagel, T. 1979. The fragmentation of value. In *Mortal Questions*. Oxford: Oxford University Press.

Offe, C. 1984. The divorce between form and substance in liberal democracy. In *Contradictions of the Welfare State*. John Keane, ed. Cambridge: MIT Press.

Offe, C. 1985. Reflections on the welfare state and the future of socialism: An interview. In *Contradictions of the Welfare State*, John Keane, ed. Cambridge: MIT Press.

Parfit, D. 1984. *Reasons and Persons*. Oxford: Oxford University Press.

Peet, J.R. 1975. Inequality and poverty: a Marxist-geographic theory. *Annals, Association of American Geographers* 65: 564–571.

Pigou, A.C. 1932. *The Economics of Welfare*. 4th edn. London: Macmillan.

Pocock, J.G.A. 1975. *The Machiavellian Moment: Florentine Political Thought and the Atlantic Republican Tradition*. Princeton: Princeton University Press.

Polanyi, M. 1962. The republic of science. *Minerva*, 1: 1–72.

Quade, E.S. 1970. Why policy sciences? *Policy Sciences*, 1: 1–2.

Sartre, J.-P. 1955. Dirty hands, in *No Exit and Three Other Plays*. New York: Vintage.

Silva, E.T., and Slaughter, S. 1984. *Serving Power: The Making of the Academic Social Science Expert*. Westport: Greenwood Press.

Smith, D.M. 1971. Radical geography – the next revolution. *Area* 3: 153–157.

Soja, E.W. 1987. The post-modernization of geography, *Annals, Association of American Geographers* 77: 289–294.

Stamp, L.D. 1960. *Applied Geography*. Harmondsworth, Middlesex: Penguin.

Stoddart, D.R. 1986. *On Geography: And Its History*. Oxford: Blackwell.

Toulmin, S. 1966. The complexity of scientific choice II. *Minerva*, 4: 155–169.

Walzer, M. 1976. Political action: the problem of dirty hands. *Philosophy and Public Affairs* 2: 160–180.

Williams, B. 1985. *Ethics and the Limits of Philosophy*. Cambridge: Harvard.

Woodward, J. 1982. Paternalism and justification. *Canadian Journal of Philosophy* 8, suppl., 67–89.

R.G. Stubbings
Department of Political Sciences and
School of Public and Environmental Affairs
Indiana University
Bloomington, IN 47405
U.S.A.

Kingsley E. Haynes
Department of Geography and
Institute of Public Policy
Boston University
Boston, MA 02215
U.S.A.

SECTION II

The Education Issue

3. Applied Geography: An Academic Response to the Structural Change of a Discipline

The Applied Geography Movement

Definitions of Applied Geography

The first published reference to applied geography has been traced to British geographer John Scott Keltie's book, *Applied Geography: A Preliminary Sketch*, which first appeared in 1890 (see Dunbar 1978; Stevens 1921). Keltie intended to show the importance of geographical knowledge on human interest, history, and especially industry, commerce, and colonialization (Keltie 1908). Shortly thereafter a committee to recommend changes in the curricula of American secondary school education defined applied geography as the study of geographical elements as they appeared in other fields of study – i.e., as an element in the study of history or botany (U.S. Bureau of Education 1893).

More recently there has again been considerable debate, as well as ambiguity, over an appropriate definition. The term has been used to refer to that 'practical knowledge' useful in everyday life, or necessary to understand the world to improve the quality of life. A few reject the term as meaningless (see survey results in Harrison 1977a); others feel its use is 'unfortunate,' implying as it might '...the existence of a nonapplied, or inapplicable, geography, or perhaps even a useless geography' (Dunbar 1978, p. 239). More often, applied geography has been linked or confined to distinct branches of the discipline, especially to economic (Herbertson 1900) and business geography (Applebaum 1956), cartography (White 1893), or, more recently, to the professional fields of planning and environmental management.

The tie between geography and planning has been, and in many instances continues to be, especially strong (Harrison and Larsen 1977). The view that 'town planning is the art of which geography is the science' (see Lever 1974, p. 58) has been argued, especially in British town planning. This obviously helped to serve as a rationale for the nearly one-half of all geography departments that indicated that they either had added or intended to add planning programs or coursework (Harrison and Larsen 1976). Some have essentially equated the fields: '...spatial planning, or urban and regional planning, is essentially human geography, harnessed or applied to the positive task of action to achieve a specific objective' (Hall 1975, p. 5). More recently, Kenzer has ventured a similar view – i.e., that there may be no distinction between human and applied geography (Kenzer 1984).

The British journal *Applied Geography* has been published since 1984. The journal's purpose statement limits its scope to 'research into man's evaluation,

M. S. Kenzer (ed.), Applied Geography: Issues, Questions, and Concerns, 37–52.
© 1989 *Kluwer Academic Publishers.*

exploitation and management of the world's resources,' arguing that this
'encompasses aspects of both human and physical geography, as well as parts of
agriculture, ecology, planning and politics.' The dominant view, however, does not
confine the scope of applied geography to a particular subtopic or skill area. Rather,
it is regarded as an extension and an application of the science of geography – i.e.,
the 'practical' use of 'pure' theoretical research. An analogy might be that theoretical geography is to applied geography what scientific medical research is to clinical
medicine. In like manner, theoretical geography seeks to describe and better
understand the world; applied geography seeks to use the content and method of
theoretical geography to aid in the resolution of world problems.

Applied geography also differs from theoretical geography in that it is (1) user-
oriented, (2) action-oriented, and (3) extends the scientific method to include
evaluation and implementation stages (Frazier 1982, p. 14). It must also be
involved in the formulation of goals and strategies and engage in value judgments,
a matter of considerable concern in the application of any scientific field. It seems
likely that many geographers would agree that both theoretical and applied
orientations are needed if geography is to have a significant impact on contemporary issues.

The debate over applied geography is likely to continue, especially within the
academic community. In the meantime, the consensus seems to be that the crux of
applied geography involves problem-solving (Frazier 1978; Harrison 1977a). In the
broadest sense, then, applied geography is defined in this paper as *the use of
geographic content, principles, and methods in research and other activities
designed to aid in the resolution of human problems with a strong geographic
dimension*. Applied geography is a mixture of 'applicable' and 'applied' geography; it includes research that has the potential for being utilized by others as
well as research performed for specific clients (Frazier 1986). In a survey of geographers, nearly 70 percent argued for such a definition, although some would
confine the aspects of geography to be applied or the types of problems to be
solved (Harrison 1977a) – e.g., the use of geographical knowledge within the
decision-making process, most notably to advance public policy (Berry 1972).

American Geography's Applied Past

The pragmatic foundations of geography can, of course, be traced to the earliest
geographic thinking. Geographic information and method have often been utilized
for practical gain or to address some problem of the time. Applied geography also
has a broadly based place in the heritage of American geography (James 1972;
Frazier 1982). A glimpse of this heritage is noted in the early efforts of the
American Geographical Society, which, at the turn of the twentieth century, had as
one of its purposes the promotion of business interests of practical use to the
merchant or missionaries (Wright 1952). There was also a growing dissatisfaction
with the pure academic orientation in the university curriculum in the early 1900s,
leading many geographers to undertake applied studies, for example, in land

classification and utilization, resource management, and economic development (James 1972).

There has been a persistent view in American geography that environmental and spatial information should be used to help solve immediate problems. Charles Colby, in the 1930s, noted that 'the close association of geographic science and geographic application is central to American geography' (Colby 1936). Colby's point may be illustrated by the work of Carl Sauer and Isaiah Bowman.

Sauer was certainly a scholar; he also engaged in applied geography. His work, for example, in developing a land classification scheme for the state of Michigan or his participation in the actual implementation of a program to investigate climatic impact on the relationship between slope and soil by the then Soil Erosion Service, certainly involved the applications of geography to real world problems (Leighly 1976). Bowman has been called America's 'most effective and well-known' applied geographer (Frazier 1982, p. 11). In addition to serving as a consultant to the federal government, notably in connection with the Versailles Peace Conference, he engaged in both foreign and domestic boundary mediations, served as a special advisor to the government on international affairs, and has been dubbed one of the 'architects of the United Nations' (Ogilvie 1950).

Many other examples could be cited. The work of Harlan Barrows in water resource management and policy development for the Public Works Administration (Colby and White 1961), or that of William Applebaum in market delineation and measurement (Epstein 1978) are notable. In wartime, geographers have applied their background and skills in logistical and military planning, cartography, and air photo interpretation. The systematic application of geographic research within the private sector, especially in locating industrial and commercial establishments, is also not new.

Decline, Evolution, and Change

While applied geography has a distinguished place in the history of American geography, its importance should not be overemphasized (Frazier 1982, p. 13). Relative to the research efforts in human and physical geography, the amount of work in applied geography has been minor. Applied research has also tended to run in cycles, rising most notably during wartime and other periods of crisis. This should be expected of any applied science; it will respond to human or national needs. Usually the strongest cries for relevance in any discipline, geography included, have come when that field fails to respond to the pressing issues of the day.

The argument that applied geography sits apart from the tradition of American geography can be traced to the dominance during the 1960s of (1) the fervor to produce objective spatial theory – the so-called logical positivistic paradigm – and (2) the need to train more teachers. These two directions so completely dominated academic geography during this time that any interest in or demand for applied geography went largely unrewarded or unnoticed. By the late 1970s, however,

social and environmental concerns at the national level, aggravated by the tensions
of Vietnam, resulted in renewed demands for relevance within geography, both in
research and curricula. Positivism, while it enhanced scientific rigor and quantitative methodology, was seen by many as largely irrelevant for solving human
problems (Frazier 1978).

During this same time geography was severely impacted by a deteriorating
educational market caused by falling birth rates, population migrations resulting in
regional declines – especially in some of the traditionally strong centers for
academic geography, most notably the American Midwest – and general academic
and economic retrenchment. Within a few short years traditional academic
geography was in a precipitous decline, as reflected by over a 40 percent reduction
in the number of academic units offering geography coursework and in excess of
30 percent losses in the number of faculty positions and students enrolled in
college-level geography courses for the period 1970–1986 (Table 1).[1]

Table 1. Seventeen-year summary of national geography enrollments and number of faculty
employed, 1970–1986

Year	Enrollment (in thousands)	Institutions where geography is taught	Number of faculty
1970	787	1260	3735
1971	789	1305	3784
1972	747	1111	3465
1973	734	1159	3519
1974	696	1174	3370
1975	695	1200	3435
1976	665	1221	3457
1977	666	1262	3471
1978	690	1343	3561
1979	671	1388	3377
1980	539	839	2399
1981	569	921	2521
1982	588	984	2659
1983	587	1039	2779
1984	465	648	2304
1985	526	760	2507
1986	513	725	2467

Source: *Schwendeman's Directory of College Geography of the United States*. Edited by
D.R. Monsebroten. Vols. 22–39 (1971–1987). Richmond, Kentucky: Eastern Kentucky
University.

Although there was (and is) much talk about what should be done within the
profession, the response to this situation was largely *de facto* by individual
geographers. The most common early defense was to argue for geography's
relatively strong position within the rapidly expanding profession of city planning.

In a survey of four-year, college-level geography departments, it was found that nearly 50 percent reported offering some form of planning-oriented coursework (Harrison, 1977b).

When national concern turned from the urban to the environmental crisis, especially following the creation of the Environmental Protection Agency and subsequent federal legislation for clean air, water, and waste management, a similar interface was developed between geography and the developing professional field of environmental management. Geography had the advantage of not only being on the ground floor in terms of involvement, but of being able to relate scientific information, especially in earth and atmospheric science, to the professional planning and decision-making framework of the real world. While this job market expanded slowly at first, especially in terms of positions at the local level, environmental management is now institutionalized at all levels of government and private consultant activity.

Structural Change in Geography

Today's applied geography movement in the U.S. goes well beyond and is quite different from past periods of heightened interest. Our contention is that the current emphasis on applied geography is, in fact, an evolutionary change within the discipline, not a fad of passing interest. While some of the reasons for this change are similar to those that have come before, many are not. Taken in combination they represent a rapidly emerging structural change within the discipline. Although the applied geography paradigm will undoubtedly evolve and change, several of the reasons for its emergence and on-going growth can be cited. The following are perhaps the six most important reasons.

1. There is both political and institutional demand for the social sciences to contribute to the solution of real world problems. Most professional geographers recognize this educational concern for the visibility, relevance, and practicality of geography in addressing socio-economic, political, and environmental issues.

2. The movement has resulted in major changes in geography curricula. A 1977 report found that nearly 90 percent of the geography departments sampled predicted at least some additional emphasis placed on applied geography in the future; over 50 percent anticipated much more emphasis (Harrison 1977a). These predictions have been acted upon. Evidence for this comes from a cursory review of current course and curricula offerings within American geography major programs, by the rise of applied specialty groups within the Association of American Geographers (AAG), and by the increasing number of faculty with applied teaching and research interests (Russell 1983; AAG 1987). These curricula changes will not be easily reversed, especially on campuses where enrollment has been stimulated by such an orientation.

3. Geography has either developed, or borrowed and made applicable, a host of skills and methodologies that are in high demand by government and business.

An increased emphasis on rigorous training in these skill areas, together with experiential learning through internships, are the primary factors that give geography graduates an 'edge' in the job market.

4. The majority of American college students continue to have strong career orientations when selecting academic areas of study. While students – and their parents who are often paying the bill – demand and should receive a quality general education, they also demand and expect to graduate with a marketable education. The ever-present view (throughout academia) that these are somehow mutually exclusive or, worse, incompatible goals, continues to threaten the ability of geographers to develop programs that encompass both sound academic and applied coursework. Many departments find that both can be accomplished within a four-year undergraduate program; they are often complementary. Again, the argument that geography, and geographers, must be confined to the pursuit of pure research and scholarship appears to be primarily a modern notion.

   If geography is to be viable, it must be – and be perceived as by both institutional administrators and students – an important discipline offering valuable coursework. Fortunately, students are often attracted to coursework that is perceived as having strong applications, in both content and method, in dealing with real world problems. Applied geography, based upon the research and scholarship of theoretical geography, has a distinct advantage in gaining such student interest. Even during periods of enrollment decline, students have commented that they 'liked geography,' but that they could not justify additional hours at the expense of more 'relevant' courses in getting a job. Where this interest has been consistently translated into employment upon graduation, then – at most institutions – the attention and support of both students and administration is assured.

5. Geography must now compete for the student's attention and dollars within a much more diverse university curricula than that which existed in the past. The reality of the academic environment has changed, and most of what has changed since World War II has been toward specialized and applied areas. Although there are 'back-to-the-basics' movements as well as efforts to standardize curricula, the university is largely a marketplace where the student as consumer is the final judge of a subject's importance. Courses and curricula with chronic low enrollments receive little sympathy from most academic administrators or state boards of education.

6. Where applied geography gains a foothold it becomes institutionalized in the local and regional job markets. Geography programs that consistently turn out well-trained and marketable graduates set in motion an expanding network. Their graduates get good jobs within their interest area; they impress their employer and advance rapidly; the employer then seeks additional graduates from the program as positions open. This network can become institutional-ized, that is, 'built-in' over time, as these graduates advance and reach positions of influence.

Geography has reached a level of maturity that allows it to respond to world and domestic problems. The time is also right for applied geography. Many of the long held topical interests of American geographers – e.g., human-environment relations, location and market analysis, water, transportation, and population problems – are now of major, even critical, concern to the general population and political leadership. Many of the world events and crises – some of which may threaten our very survival, such as boundary disputes, starvation, acid rain, and toxic wastes – are problems with major geographic dimensions that must be resolved.

Response to Structural Change

Academic geography departments have responded to this structural change in many ways. Some chose to ignore it, labeling applied geography a 'fad' or temporary disciplinary preoccupation. More common are the departments that modified existing courses and increased demands for skill training among majors and graduate students, to include statistics, remote sensing, computer programming, regional economics, geographic information systems (GIS), computer graphics, and mapping. These departments typically advise students to buttress their academic training in geography with work in other areas, as in planning, biology, economics, marketing, computer science, public administration, or transportation.

There have been a host of program changes. New degrees, with such titles as Environmental Studies, Area Development, and Resource Management, have been added to describe more accurately the kind of education and training that geography students were receiving. Faculty training and expertise have expanded into more specialized and applied fields.

Some departments went further by adding new courses, especially in the skill areas, such as GIS. Others reported on such diverse developments as applied geography courses for transportation planners (Stutz and Heiges 1983), consulting on United Nations projects (Karan 1985), applied cultural geography (Lanegran 1986), co-op and research agreements between university geography departments, business, and government (see Lier 1984; Marotz 1983; Munn 1980; Stutz 1980; Russell 1977), and community service extension programs, including those developed through the cooperative extension service at land grant universities (Beck 1984; Ludwig 1984; Janke 1984).

Considerable concern has also been voiced for stronger support and linkages between applied geography and the nonacademic job market (Smith and Hiltner 1983). One study found that over 75 percent of the nonacademic business geographers surveyed in the Washington, D.C. area were working for engineering, environmental assessment/remote sensing, and computer mapping/GIS firms (Monte 1983). Anderson (1979) detailed some of the major government employment areas for geographers at both entry- and policy-level positions.

The Southwest Texas Model

The Southwest Texas State University (SWT) geography department had the advantage of being created rather late, 1965; planning was added to the departmental name in 1976. Thus, departmental direction was guided by early evidence of the structural change in geography that led to the rise in importance of applied geography. The department's movement toward applied geography was also aided by a set of circumstances that may be unique to SWT. Southwest Texas is a large (11,800 in 1972; 20,000 in 1987) regional university with undergraduate and masters programs. Surprisingly, for such a large campus, there are no separate programs in geology, planning, environmental design, architecture, or resource management. The Business School has had little emphasis on regional economics or on the spatial aspects of marketing; nor is there a school or department of engineering. Finally, the School of Science (and more particularly the Department of Biology) has not chosen to assume leadership in the development of a human-environmental studies program.

This local situation, combined with the evolving structural change in geography, largely dictated the nature of the academic program at SWT. In fact, one could claim that had the department not changed, it may not have survived. By 1971, the department was offering B.A. and B.S. degrees with concentrations in urban and regional planning, resource and environmental studies, and cartography and photogrammetry. There was also the B.S. degree in education with a first teaching field in geography. The department experienced slow growth between 1971 and 1977. Meanwhile the nation experienced the Arab oil embargo, a continuation of the urban crisis, and acute environmental problems. Changes in the basic fabric of American society accelerated structural change in geography. Problems that had been interesting research topics for geographers became driving concerns for the country. The demand for college-educated people who could understand the delicate balance between the human and natural environment exploded. Simultaneously, the American college student changed too. Relevance gave way to practicality and the need for a high paying job upon graduation. Young, upwardly mobile professionals (YUPPIES) turned away from history, philosophy, and English, demanding entrance into business schools and other applied disciplines.

The SWT departmental response in the late 1970s was to fine-tune its curriculum and expand its internship program. A job placement assistance program was initiated that focused on a forty-page booklet: 'How to Get a Job in Geography.' Numerous handouts were prepared alerting students to the job opportunities for those who would graduate with degrees in geography.

More importantly, our graduates were getting jobs. In 1982, a survey of sixty-three recent graduates showed that forty-one held positions where they were using their geographic training; the demand was particularly oriented toward geographic skills. The job successes of our graduates from 1977 onward trickled back to campus, creating demand for geographic training that has reached almost unmanageable proportions. Figure 1 shows growth in total annual enrollment from about 1500 in 1978–79 to nearly 4500 in 1986–87. The 1987–88 enrollment was

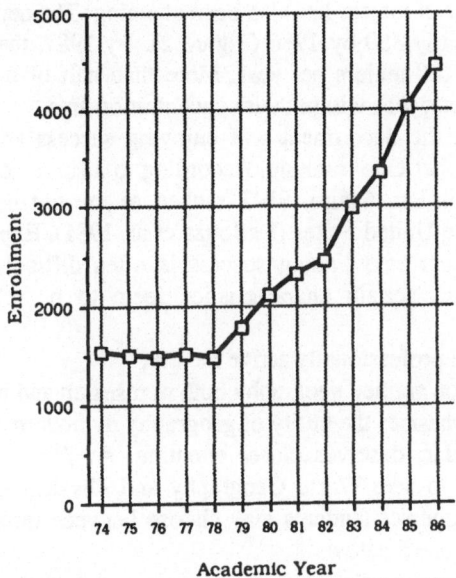

Fig. 1. Total Enrollment, 1974–87, Department of Geography and Planning, Southwest Texas State University.

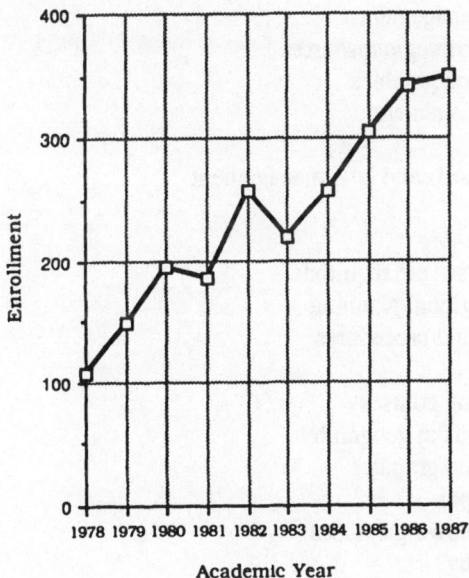

Fig. 2. Undergraduate Majors, 1978–87, Department of Geography and Planning, Southwest Texas State University.

4633. Growth in the number of majors has also been stunning. The department had 110 majors in 1978 and nearly 350 by 1986 (Figure 2). By 1987, the department was graduating more than 100 majors per year. More than half of these students have been successful at getting jobs within their concentration area.

In 1981, assurances that the department was enjoying success and prosperity came in the form of a 'Number One' ranking. According to a survey carried out at the University of Wisconsin-Eau Claire, SWT ranked as the top undergraduate geography department in the United States (DeSouza et al. 1981). Explaining such a ranking and the department's continuing success is often difficult because of intangible factors. However, certain characteristics seem to be related to the department's prosperity:

1. a relatively young and professionally active faculty;
2. a departmental focus on applied geography both in research and in teaching;
3. a curriculum that emphasizes the utility of geography in modern society.

The curriculum, in particular, deserves closer attention. At SWT every under-graduate major is required to take World Geography and Physical Environment. Once they complete these courses students may choose between three concentrations; each concentration is listed below.

Resource and environmental studies
3301 Quantitative methods in geography
3303 Economic geography
3313 Natural resource use and planning
3321 Energy resource management
3334 Water resources management
3343 Global patterns of strategic resources
3411 Map compilation and graphics
4313 Environmental management
4338 Land use planning
4339 Environmental hazards and land management
4380 Internship

Urban and regional planning
The following three courses are required:
3320 Community and regional planning
4321 Planning methods and procedures
4338 Land use planning
Select one of the following courses:
3301 Quantitative methods in geography
3411 Map compilation and graphics
4422 Computer cartography
Select any four of the following courses:
3303 Economic geography
3310 Urban geography
3313 Natural resource use and planning
3321 Energy resource management
3323 Location analysis

3334 Water resources management
4313 Environmental management
4336 Transportation systems
4350 Solid waste disposal planning and management
4380 Internship
Cartography and GIS
3301 Quantitative methods in geography
3411 Map compilation and graphics
3415 Cartographic production
3416 Air photo interpretation
4412 Remote sensing
4422 Computer cartography
4430 Field methods
4440 Geographic information systems

These courses suggest geography's important advantage: the core of the discipline includes matters of critical concern to society as a whole. Consequently, employers want and need young people who understand human-environmental relations, transportation, planning, mapping, location, movement, and the complexities of various types of regions. For perhaps the first time in recent history, geography seems to be the envy of other academic disciplines.

Masters of Applied Geography

Clearly, the SWT geography program has focused its efforts on applied geography. Faculty hirings and curricular change have strengthened this emphasis. When departmental planning began in 1981 for a graduate degree, the senior faculty sought to build on this orientation by developing the first Masters of Applied Geography (M.A.G.) in the United States. The M.A.G. degree, first offered in 1983, requires thirty-nine hours with a nine-hour core including courses in quantitative methods, research design and techniques, and applied geographic analysis. A fifteen-hour concentration may be taken in one of three areas: physical and environmental studies; land/area development and management; cartography and GIS. Six hours must be taken in an outside discipline. The exit module consists of a six-hour thesis and a three-hour internship, or a six-hour internship combined with three hours of directed research.

The M.A.G. curriculum
Required core (9 hours)
5300 Research design and techniques
5301 Quantitative methods
5309 Applied geographic analysis
15 Hour concentrations
A. Physical and environmental studies
    5313 Environmental management
    5314 Geographic elements of environmental law

5316 Applied physical geography
plus 6 hours of geography electives
B. Land/area development and management
5312 The planning function and process
5338 Land use planning
5339 Land development and management
plus 6 hours of geography electives
C. Cartography and GIS
5408 Applied cartography
5416 Remote sensing
5417 Computer cartography
5418 Geographic information systems
Electives:
5315 Regional analysis
5323 Locational analysis
5335 Directed research
5336 Transportation systems
5337 Impact assessment of land development
5340 Geography for teachers
5351 Regional waste management
5360 Seminar in planning problems
5370 Seminar in applied physical geography
5395 Problems in applied geography
5430 Field methods

The M.A.G. degree parallels the success of the undergraduate program. There were forty-eight students in 1987. About one-half were full-time students while the remainder worked and attended school at night. Approximately one-third were SWT undergraduates, one-third were graduates from other Texas universities, and one-third were from out-of-state.

Thus far, there are two very specific measures of success. There have been ten M.A.G. graduates. Nine have relatively high-paying, mid-level positions in areas related to their concentration. The tenth student is preparing for Ph.D. work. The other measure is that a second M.A.G. degree has come into existence (1987) at New Mexico State University. Their degree is somewhat different from ours but nonetheless was created to serve many of the same demands that led to the initiation of the SWT degree.

Adjustments in the SWT Model

Degree programs in applied geography must necessarily be flexible. To some extent, demand, or the need for education and training, is dependent on tendencies of federal funding, local and regional growth patterns, and environmental concerns. In Texas, for example, there is growing belief that 'water, not oil, is the resource of the future.' Another imperative is the federal and state concern over the handling and storage of hazardous waste materials, a problem with considerable geographic dimensions. With these and related concerns, rational planning requires an abundance of reliable data in easily retrievable form.

These factors are reflected in the most recent curricula changes in the SWT program. At the undergraduate level, new courses have been added in Water Resources Management, GIS, and Computer Mapping. At the graduate level, new courses are available in Impact Assessment of Land Development, GIS, and Regional Waste Management; Geographic Aspects of Environmental Law is now offered more regularly.

We have also accelerated the offering of Land Use Planning, Locational Analysis and Environmental Management, while decelerating courses in regional geography and energy management. The internship program at both the undergraduate and graduate levels has been expanded dramatically. Currently, approximately twenty-five students are interns in governmental and private agencies each semester, including summer. Some interns are paid; all receive 3–6 hours of credit for between 150–300 hours of work. In the last four years, there have been more agency requests for interns than the department has qualified students to fill them.

Adjustment of the SWT Model for Use at Other Universities

Applied geography is not the only answer to the problem of declining enrollments and disappearing departments, nor is it recommended as a major thrust for all departments. It is, however, a strategy that often makes sense to students, university administrators, employers in the public and private sectors, and to parents. It is our job as professional geographers to recognize that geography has great utility and to pass it on to students in the form of contemporary education and training. To ignore the application of geography is to do a disservice to the profession and to those students who come to the field for academic training.

Some departments have created programs of applied geography well suited to the university and the external environment in which they are situated. One of the first to report developing an applied undergraduate program was Ryerson Polytechnical Institute in Ontario (Foster and Jones 1977). Other notable departments include:

1. Salem State College, with a vigorous program in travel and tourism, which seems particularly appropriate in the Boston/New England area;
2. Southwest Missouri State University, with both undergraduate and graduate concentrations in resource planning;
3. University of Idaho, with programs in physical environment and earth resources, mineral property and land management, and applied economic geography;
4. University of South Carolina, with undergraduate and graduate programs in applied geography and GIS;
5. New Mexico State University, with undergraduate and graduate programs in computer cartography, remote sensing, land use planning, environmental

planning and community development;

6. State University of New York-Binghamton, where two of the graduate-level tracks provide a focus on applied geography: 'Analytical Geography' includes most of the skill courses, internships, and preparation for the Ph.D.; 'Geography and Planning' emphasizes urban analysis and environmental management. The department publishes annually *Proceedings of the Applied Geography Conference* and *Research in Contemporary and Applied Geography.*

7. East Carolina University, with a formal blending of geography and planning. The department offers the B.S. in Urban and Regional Planning. Practical experience is encouraged through internships and cooperative education.

These seven departments do not represent a complete list; they are those with which the authors are familiar. However, they are all successful and they enjoy considerable respect locally. The faculty are involved in applied geographic research in their communities. More importantly, they are educating students with the message that geography has great utility, and their graduates now occupy key positions in local and statewide job networks. In turn, many of the graduates are hiring additional geography majors, helping to insure the discipline's continuing health and vitality.

Obviously, applied geography is not the right path for all departments. Still, it is likely that a careful analysis of the local university environment will uncover niches that could be filled by an aggressive geography department. Certainly one of geography's great strengths is its eclecticism.

For those programs considering an applied geography option, we recommend that departmental planning groups first examine the local and regional environment. What priorities are apparent? What areas are expanding? What are the needs for people with geographic training? It is also necessary to assess the educational background and experience of existing faculty. Do they have the interest and energy to support program development in applied geography? Do they have the necessary skills to respond to the needs of the surrounding area? Further, a careful assessment of existing academic units on the campus is required to include what those units do. For example, what aspects of environmental management or community planning are not controlled by another unit? Who teaches locational analysis or the spatial aspects of marketing? Who teaches computer mapping or computer graphics?

Once this and kindred information has been collected, logical decisions about curricula changes can be made. These changes may be as minor as reorganizing course content or as great as suggesting new courses to be added to the inventory. It may be that the time is right for a new degree or to change the name of an existing one.

Note

1. The problems of using the *Schwendeman's Directories of College Geography* to collect annual departmental data are evident, especially when any two consecutive years are compared; yet, the overall trend in Table 1 is inescapable. One way to 'smooth' the abrupt differences from year to year – which are caused mainly by holding information about a program for three years if it is not updated by the department – is to average, say, any three years. In Table 1 this was done for the two periods: 1970–72 and 1984–86 (current). The declines can then be measured: a loss of 273,000 students or a 35.3 percent decline in enrollments; a 42.0 percent decline in the number of institutions where geography is taught; and a loss of over 1235 (33.7 percent) faculty positions in geography.

References

Anderson, J.R. 1979. Geographers in government. *Professional Geographer* 31:265–270.

Applebaum, W. 1956. What are geographers doing in business? *Professional Geographer* 8:2–4.

Association of American Geographers. 1987. *Guide to Departments of Geography in the United States and Canada 1987–1988.* Washington, D.C.

Beck, R.C. 1984. Opportunities for geographers in the cooperative extension service. *Professional Geographer* 36:234–235.

Berry, B.J.L. 1972. More relevance and policy analysis. *Area* 4:77–80.

Colby, C.C. 1936. Changing currents of geographic thought in America. *Annals, Association of American Geographers* 26:1–37.

Colby, C.C., and White, G.F. 1961. Harlan Barrows, 1877–1960. *Annals, Association of American Geographers* 51:395–400.

DeSouza, A. *et al.* 1981. The overlooked departments of geography. *Journal of Geography* 80:170–175.

Dunbar, G.S. 1978. What *was* applied geography? *Professional Geographer* 30:238–39.

Epstein, B.J. 1978. Marketing geography: a chronical of 45 years. *Proceedings, Applied Geography Conference* 1:373–380.

Foster, L.T., and Jones, K.G. 1977. Applied geography: an educational alternative. *Professional Geographer* 29:300–304.

Frazier, J.W. 1978. On the emergence of an applied geography. *Professional Geographer* 30:233–237.

Frazier, J.W. 1982. Applied geography: a perspective. In *Applied Geography: Selected Perspectives,* edited by J.W. Frazier *et al.,* pp. 3–22. Englewood Cliffs, NJ: Prentice-Hall.

Frazier, J.W. *et al.,* 1986. *Papers and Proceedings, Applied Geography Conferences* 9:xvii.

Hall, P. 1975. *Urban and Regional Planning.* New York: John Wiley and Sons.

Harrison, J.D. 1977a. What *is* applied geography? *Professional Geographer.* 29:297–300.

Harrison, J.D. 1977b. Geography and planning: convenient relationship or necessary marriage? *Geographical Survey* 6:11–24.

Harrison, J.D., and Larsen, R.D. 1976. Planning curricula in departments of geography. *Bulletin, Association of Collegiate Schools of Planning* 14:1–5.

Harrison, J.D., and Larsen, R.D. 1977. Geography and planning: the need for an applied geography interface. *Professional Geographer* 29:139–147.

Herbertson, A.J. 1900. Applied (or economic) geography. In *Report of the Commissioner of Education for the Year* 1898–99, 1:1189–1208. Washington: Government Printing Office.

James, P.E. 1972. *All Possible Worlds*. Indianapolis: Odyssey Press.

Janke, J. 1984. Geography in university cooperative extension. *Professional Geographer* 36:240–241.

Karan, P.P. 1985. Geographers as consultants on UN projects. *Professional Geographer* 37:470–473.

Keltie, J.S. 1890, rev. 1908. *Applied Geography: A Preliminary Sketch*. London: G. Philip and Son.

Kenzer, M.S. 1984. Comments from the outside: the sixth annual applied geography conference, 12–15 October, 1983. *Applied Geography*. 4:85–86.

Lanegran, D.A. 1986. Enhancing and using a sense of place within urban areas: a role for applied geography. *Professional Geographer* 38:224–228.

Leighly, J. 1976. Carl Ortwin Sauer, 1889–1975. *Annals, Association of American Geographers* 66:337–348.

Lever, W.F. 1974. Geography and planning. In *Studies in Social Science and Planning*, edited by J. Forbes, pp. 55–80. New York: John Wiley and Sons.

Lier, J. 1984. Comments on industry-academic cooperation. *Professional Geographer* 36:219–221.

Ludwig, G.S. 1984. Extension geography: applied geography in action. *Professional Geographer* 36:236–238.

Marotz, G.A. 1983. Industry-government-academic cooperation: possible benefits for geography. *Professional Geographer* 35:407–415.

Monte, J.A. 1983. The job market for geographers in private industry in the Washington, D.C. area. *Professional Geographer* 35:90–94.

Munn, A.A. 1980. The role of geographers in the department of defense. *Professional Geographer* 32:361–364.

Ogilvie, A.G. 1950. Isaiah Bowman: an appreciation. *Geographical Journal* 115:229.

Russell, J.A. 1977. Modern geography: foundation of corporate strategy. *Professional Geographer* 29:200–207.

Russell, J.A. 1983. Specialty fields of applied geographers. *Professional Geographer* 35:471–475.

Schwendeman's Directory of College Geography of the United States, edited by D.R. Monsebroten. Vols. 22–39 (1971–1987). Richmond, Kentucky: Eastern Kentucky University.

Smith, B.W., and Hiltner, J. 1983. Where non-academic geographers are employed. *Professional Geographer* 35:210–213.

Stevens, A. 1921. *An Introduction to Applied Geography*. London: Blackie and Son.

Stutz, F.P. 1980. Applied geographic research for state and local government: problems and prospects. *Professional Geographer* 32:393–399.

Stutz, F.P., and Heiges, H.E. 1983. Developing courses in applied geography for transportation planners. *Professional Geographer* 35:206–210.

United States Bureau of Education. 1893. *Report of the Committee on Secondary School Studies Appointed at the Meeting of the National Educational Association, July 9, 1892*. Washington: Government Printing Office.

White, A.S. 1893. The position of geography in the cycle of the sciences. *Geographical Journal* 2:178–179.

Wright, J.K. 1952. *Geography in the Making*. New York: American Geographical Society.

Richard G. Boehm and James D. Harrison
Department of Geography and Planning
Southwest Texas State University
San Marcos, TX 78666–4616
U.S.A.

4. Cooperative Education and Applied Geography

Applied geography has generated much discussion within the discipline over the past decade. Among academic geographers, deleterious trends within higher education, including declining enrollments, departmental reductions and closures, and erosion of traditional labor markets, have stimulated that discussion. In some people's minds, applied geography is a solution for these inimical developments (Foster and Jones 1977; Frazier 1982).

Among the issues in the literature addressing applied geography are how to incorporate such a program into a department and its impact on the educational and research missions of academic units. The goal of this essay is to review an educational mechanism that departments can employ to facilitate its development of applied geography – cooperative education.

Purpose and Perspective

One objective of this discussion will be to develop an operational definition of applied geography and its relationship to academic geography. Another primary purpose of the essay is to review the nature of experiential education, particularly cooperative education, and to evaluate its potential contribution to the educational mission of departments. Also the curriculum ramifications of cooperative education will be addressed, as will some of the more critical faculty roles.

This exposition is developed exclusively from the perspective of an academician concerned about the contemporary problems affecting geography in higher education. Furthermore, the focus is on the instructional role of departments; the research function is largely ignored, a perspective resulting from the author's administrative function. An inherent assumption in this narrative is that experiential education is an invaluable asset to the academic curriculum. This belief is born out of twelve years involvement in experiential education, including a departmental internship and a university-wide cooperative education program.

Applied Geography: What Is It?

Considerable ambiguity surrounds the term 'applied geography,' which may be partially the result of the profession's disagreement on the core of geography (Morrill 1983). Geographers also disagree on the historical roots of applied geography. Whereas Harrison (1977) dated its origins to the 1950s, Ford (1982, p. 132) viewed it to be 'a rallying point [that] has arisen only in the last decade.' In

M. S. Kenzer (ed.), Applied Geography: Issues, Questions, and Concerns, 53–64.
© 1989 *Kluwer Academic Publishers.*

contrast, Dunbar (1978) asserted that applied geography existed in 1890, if not before.

One of the more extensive discussions of the nature and examples of applied geography is the book edited by Frazier (1982). Frazier characterized the relationship between basic and applied geography, noting that:

Applied geography uses the principles and methods of pure geography but is different in that it analyzes and evaluates real-world action and planning and seeks to implement and manipulate environmental and spatial realities. In the process, it contributes to, as well as utilizes general geography through the revelation of new relationships (1982, p. 17).

Furthermore he argued that an applied geographer joins the 'world of opinion' with the 'world of decision' (Frazier 1982, p. 17). In essence, Frazier argued that applied geography is utilizing geographic knowledge in a decision-making context and it is action-oriented.

Beaujeu-Garnier included applied geography as one component of the discipline that 'is essentially an attitude of mind and that its purpose is to influence or even change the future' (Beaujeu-Garnier 1975, p. 280). She further differentiated between applied geographic research focused on a specific problem confronting an organization versus pure or basic research which may only have implications for the 'practical domain.'

Harrison (1977) conducted a survey of geography departments to develop a consensus definition of applied geography. He reported: 'The majority view interprets applied geography as the use of geographic methods and theory in problem solving' (p. 299).

Applied Geography and Geographers

Based on the preceding ideas, applied geography in this discussion will be defined as problem solving or decision-making using geographic knowledge. Furthermore, it entails decision-making and problem solving in the 'practical domain.' Given this definition, applied geographers are those persons who either solve problems, make decisions, or implement the results of those decisions while using geographic knowledge. Obviously, nonacademics employed in the private and public sectors produce the great majority of applied geographic work. The role of academic geographers in applied geography is less evident because as Hart (1978) suggested, few academicians, geographers or otherwise, play a significant role in decision-making.

Role of Educators

Academic geographers can contribute to applied geography through various activities, research being the most obvious. However, much of the geographic literature would fall into Beaujeu-Garnier's category of only having implications in the real world instead of being applied geography. In contrast, those academic geographers who perform consulting services for private and public organizations will more likely shape decisions, if not make them (Stutz 1980). Utilizing one's geographic expertise in local community issues is another possibility (Jumper 1975). Community involvement through the cooperative extension service represents an applied avenue for academics (Bein 1984). Not to be overlooked is continuing education for professionals in allied fields (Bein 1984; Stutz and Heiges 1983). This option will become more attractive given higher education's growing emphasis on nontraditional students.

The most significant contribution of college professors to the development of applied geography is undoubtedly their role as instructors. As Hart (1978, p. 253) correctly reasoned, educators 'have the responsibility and pleasure of trying to help our students learn how to make better personal decisions.' One of the more important personal decisions that a student makes is an informed career decision.

In essence, most applied geography is performed by nonacademics in the public or private sectors. Although educators can contribute indirectly to the field, it is the author's opinion that their primary role is the training of students to be applied geographers. Cooperative education acts as the link between applied geography and educational programs.

Cooperative Education: Definition and Value

Cooperative education and internships are two forms of experiential education that provide students with on-the-job experience to supplement and complement their classroom endeavors. Numerous people have discussed the nature of geography internships (Gad 1979; Heiges 1972; Heiges 1977). In contrast, cooperative education has received less attention (Foster 1982; Spinelli and Smith 1981). In this section, the nature of cooperative education will be described and compared with internships. Also the value of co-oping to students and departments will be considered.

Definition

Although cooperative education is similar to internships (in terms of providing students with practical experience), it does possess some distinctive characteristics. Most importantly, cooperative education requires an alternation of work and study. Undergraduate students will alternate academic terms between college and work from the end of their freshman or sophomore year until graduation. In contrast, graduate students may complete only two work assignments. Some universities use the 'parallel' model in which a student alternates by daily working part time and attending classes part time. The alternation feature and multi-term commitment allows for greater integration of academic and work programs than do internships, which usually encompass only one academic term.

A second distinctive feature of cooperative education is the requirement that employers pay co-op students, whereas interns may be unpaid. Furthermore, the co-op student's work assignments will entail greater responsibility and challenge as he or she progresses through college. Normally interns will perform the same level of duties throughout their work assignment.

In 1985, 635 institutions of higher education in the United States had cooperative education programs and in 35 of those institutions geography was one of the participating disciplines (Cooperative Education Association 1986). Geography's role in cooperative education is clearly growing, because only 14 departments offered the co-op option in 1978 (Cooperative Education Association 1978).

Advantages to the Student

Cooperative education and internships possess some similar educational advantages for students. Students are able to apply their geographic training to 'real world' problems, observe the operations of nonacademic employers, identify career opportunities, and test career objectives. Improvement of communication skills is a frequent outcome of these experiences. Graduate students may be able to integrate their work assignments with thesis or dissertation requirements, which will strengthen both the academic and practical components of the research. Also, students report that classroom activities are more meaningful after work assignments because a focus is provided for their educational programs.

Not to be overlooked is the fact that these students have a distinct edge in the labor market because they acquire the experience to surmount some of the employment obstacles faced by most students, particularly those in the liberal arts. These barriers include poorly defined career goals, ignorance of basic business ideas, an inability to see how their training can benefit prospective employers, and a lack of self-confidence (Ritchie 1983).

These contributions to students' educations are shared by cooperative education and internships. However, the intensity of such benefits is often greater for co-ops than interns because of the planned progression in co-op work responsibilities and co-op's greater time commitment. One unique advantage of cooperative education is that it normally entails substantial on-the-job training, which may even include formal classroom training at the work site. For example, one geography graduate student spent four weeks being trained on his employer's mainframe and personal computers, including JCL, software, and programming languages. Of particular value was exposure to the company's geographic information system. The training component of cooperative education will be of greater magnitude than in internships due to the shorter duration of internships.

Another reason cooperative education involves more training is that co-op employers use it as a training and recruiting mechanism. Many employers hire co-op students specifically to train them, assess their on-the-job performance, and recruit them as employees after graduation if their performance is satisfactory. Normally the recruitment commitment by co-op employers is greater than in the case of intern employers, resulting in their rate of job offers being higher as well. Since employers have a strong interest in the success of co-op students, they strive to integrate their co-op students more fully into the organization than are interns, particularly unpaid interns.

On-the-job training can be an important adjunct to geography curricula. The training is indepth and oriented to the needs of a specific industry and it provides a mechanism for exposing students to the specialized skills sought by prospective employers without continuously having to adjust curricula to labor market fluctuations. Also it is typically of high quality because, as Beard (1976) and Richardson (1986) observed, nonacademic geographers often have the financial resources to utilize more sophisticated research tools and more extensive original data bases than do academics. In general, employer-sponsored training is a specialized complement to the more generic training available on campus.

The pay for co-ops, which has ranged from $640 to $1800 per month at Bowling Green State University, is clearly an asset. It facilitates students relocating to nonlocal work sites, and for a rural university most employers will not be local. Also, the income for undergraduates will offset some college expenses – an important feature in this era of declining financial aid. In the case of graduate students, co-op salaries can supplement departmental assistantship funds or provide graduate students with incomes to supplement their stipends. Furthermore, earning pay contributes to a student's sense of self-worth and confidence. Although pay may result in greater employer control over co-ops than interns, as Heiges (1977) observed, greater control does not imply detrimental outcomes. In those rare cases where the work experience is unsatisfactory, the college can halt the placement of students at the work site.

Contributions to the Discipline

The most important reason departments should consider cooperative education or internships is that they enhance students' educational programs. Furthermore, they are marketing mechanisms for the departments and the discipline. The need for geography marketing itself has been stressed by Carter and Steinbrink (1974, p. 1): 'Geographers, generally, have been more concerned with producing geographers than making the discipline respectable, relevant, and viable to the public.' Further they suggested that 'the traditional product-concept (that is, advancing geography as a body of knowledge while producing more geographers) is shortsighted and should be reoriented toward the accepted market concept that stresses clientele satisfaction, problem solving and accountability.' What better method to market the discipline than to supply our clientele (nonacademic employers) with sample products (geographic training embodied in our students)?

Some academicians have lamented the fact that geographers must compete with planners, regional economists, regional scientists, and other disciplines for jobs, with cartography being the only occupation dominated by trained geographers (Morrill 1983). Similar competition also exists in the co-op labor market. While occupational monopolies might be more satisfying, competition is a fact of life for everyone. Geographers must stop bemoaning the competition and prepare themselves to be effective competitors.

Geography faculty who monitor co-op students have numerous opportunities to visit their nonacademic counterparts. Such visitations can yield fresh materials for classes and provide input for curricula modifications. Also they can build linkages, which ultimately may lead to research and grant opportunities. In general, site visits to employers can initiate dialogues with nonacademicians who can provide fresh insights into the problems confronting the discipline (Smith and Spinelli 1979). Because co-op financially benefits employers, employers will be more receptive to educators' requests for support.

Typically, co-op students have a relatively easy time securing employment and a head start on their careers. Consequently the students feel good about their educations, which should benefit both the departments and the discipline. The labor market advantages of cooperative education can enhance graduate and undergraduate recruitment efforts since potential majors will be concerned about career opportunities. Some of these advantages are shared by both internships and cooperative education but, as noted previously, the magnitude of benefits from cooperative education is enhanced by the greater employer and student commitment.

Disadvantages of Cooperative Education

Cooperative education is not all good news; it does create some problems. One drawback of cooperative education is that it delays students' graduation dates, whereas internships are less likely to do so. Thus heavier faculty involvement is required to advise prospective co-op students about the advantages and disadvantages of co-oping, so students can make intelligent decisions regarding graduation delays. Also, careful advising is required to develop course sequencing around their work assignments. If a department has a substantial number of required courses with prerequisites, then cooperative education can present a scheduling dilemma (Foster 1982).

Another shortcoming of cooperative education (in relation to internships) is that paid work assignments will be more difficult to locate than volunteer positions, particularly in adverse economic times. There is no solution to this problem other than intensified marketing efforts. If more nonacademic geographers become active in professional organizations, then they will become key contacts to help market co-op students.

Developing cooperative education for graduate students can be problematic due to the brevity of their programs. Also, some employers, particularly in the private sector, believe graduate students are overspecialized, overtrained, and will demand unrealistic salaries. While employers' biases inhibit the development of both internships and cooperative education, internships are better suited to the timeframes of graduate programs than is cooperative education. Nonetheless, some excellent co-op opportunities do exist for graduate students, especially in the federal government, and those possibilities are worthy of exploration.

Administrative issues at some institutions may inhibit cooperative education (Foster 1982). For example, the operation of a cooperative education program requires funding, and the necessary funding may be difficult to acquire, particularly in times of tight budgets. Also, the establishment of guidelines for faculty loads and compensation for co-op work can be problematic since the nature of the work differs from the usual teaching, research, and service assignments. Furthermore, some administrators and faculty may believe cooperative education or experiential learning is not an appropriate function of a university. Such problems can constrain severely the development of cooperative education.

Even though this discussion has focused primarily on cooperative education and its advantages relative to internships, both forms of experiential education are recommended. Some students will not be able to co-op and they should have the opportunity to do an internship.

Curriculum Ramifications

The nature of curricula modifications required to strengthen applied geography is controversial. Various people have argued for substantial curricula change. For example, Stutz and Heiges (1983, p. 206) observed that 'The need to revamp academic curricula is clearly increasing as students are now requiring practical experience and a firsthand knowledge of how theories and methodologies should be applied to help solve real world problems.' In contrast, others view such trends as ill-advised and detrimental because they will result in a rejection of rich traditions in geography and ignore the liberal education mission of the discipline (Ford 1982; Salter 1976).

Liberal Arts

Nine years experience in cooperative education has proven the liberal education component of geography to be fundamentally sound. In a liberal arts program, a student should learn to accumulate and evaluate information relevant to decision making, to communicate effectively, to appreciate and evaluate the values and contributions of other cultures, to think critically and analytically, and (that uniquely geographic ability) to synthesize materials (Hart 1978; Sengenberger 1975). These attributes are valued by employers. Over the past nine years at Bowling Green State University, employers' criticisms of co-op students, both technical and liberal arts majors, have focused more on the quality of these liberal arts skills than on the co-op students' technical proficiencies. Furthermore, one often encounters statements by higher-level, corporate management reaffirming the value of liberal arts. For example, Judd Alexander, Senior Vice President of American Can, stated that 'I want imagination and organization, confidence and humility ... I want people who are smart, who know how to use their brains, and how to work well with others' (Hall 1983, p. 16). If geography neglects the teaching of the basic skills, then attempts to develop applied geography will be futile because students will not survive in the labor market.

Another justification for a solid liberal arts background is the fluctuations in the labor market. It is ludicrous to assume that a student can be trained adequately in 1989 to meet the needs of the 1999 market. Rather than constantly revamp curricula to accommodate shifts in the job market, departments should return to the 'basics' and provide students with a firm foundation for career development.

Perhaps too many faculty view an applied curriculum in the narrowest sense as vocational education. For example, Salter (1976, p. 70) argued that 'geographers should admit the "impractical" focus of their subject and celebrate it rather than apologize for it. It would be a tragedy to see the study of geography converted into an orthodox vocational exercise.' This focus on vocational education, or training for a specific job, ignores the broader perspective of career education, which is 'helping one to become aware of options. The process of becoming aware requires the liberal arts disciplines of communication and analysis' (Millonzi 1979, p. 8). Applied geography in an educational context should be viewed as career development, not vocational education; it emphasizes intellectual dexterity rather than manual dexterity.

Skills Component

Liberal arts should be the cornerstone of the curriculum of geographers, but development of applied skills cannot be ignored. Traditional geographic skills, such as field techniques, cartography, and quantitative methods, must be taught. Furthermore, these skills should be taught in a problem solving context and developed not in techniques courses alone. They should also be incorporated into all systematic and regional courses whenever possible.

Skill development is essential to facilitate students' entry into the labor market. While higher-level managers may believe that a liberal education is desirable for a business career, they actually hire few if any entry-level employees. Those in charge of hiring new college graduates focus too narrowly on candidates requiring little training and who will be immediately productive. Consequently those managers normally seek job candidates possessing skills that are immediately applicable. For this reason, the skills component should be linked to employers' needs as much as possible. As Monte (1983, p. 94) argued, 'To market geography in the business world, one should view geography in terms of the products it can produce relative to the clients who would use them.'

Acquisition of appropriate skills need not be restricted to geography courses. The skill base of students can be enlarged by the judicious selection of degree minors and elective courses. Furthermore, the on-the-job co-op training described previously also contributes to students' skill base, especially skills that hiring managers perceive to be applicable immediately. If students are successful in their entry-level positions, they will move into positions of greater responsibility. At some point in their career they will no longer be relying as much on their entry-level skills, and then their liberal arts education will assume even greater importance.

In summary, the ideal curriculum will contain both liberal arts and skill components. When cooperative education is superimposed on such a curriculum, the student will also acquire specialized training oriented toward the specific labor market of his or her choice. The exact nature of the curriculum will be determined by the departmental faculty.

Faculty Roles

The particular applied specialization of any given department will depend on the capabilities and interests of the faculty. Planning is the field most often cited as an outlet for applied geographers (Harrison and Larsen, 1977). As Frazier (1978) and others have observed, the development of applied curricula to train students for the planning market should be undertaken cautiously due to its fluctuations and the growth of professional planning schools. Also, those departments without appropriate faculty expertise should not move in that direction. Given the scarcity of funds at some universities for hiring new faculty, it is unlikely that departments can hire sufficient new faculty to reorient their programs. If the 'critical mass' for a planning specialization does not exist, then perhaps another applied focus, such as marketing geography or location analysis, can be emphasized. Another consideration should be the nature of the local labor market – i.e., the development of applied specialities will be facilitated by the availability of local employers.

With a cooperative education program, some faculty functions cannot be overlooked. One key element will be advising. Too few persons have recognized this crucial role. One exception is Hornbeck (1979, p. 47) who posed advising as an alternative to major curriculum revisions: 'Rather than responding directly to the vicissitudes of the job market by adding new courses, members of a geography department may offer a carefully organized career counseling program that emphasizes student planning as an effective alternative to a drastic alteration of current teaching programs.' While one may not agree that advising will replace programmatic modifications, it is difficult to disagree with Hornbeck about the importance of advising. Students not only need assistance through the bureaucratic maze of higher education, but also they must be appraised of career options and their personal values as related to career choices (Salter 1983). Such counseling can not be relegated entirely to the placement office and career counseling staff. Also, as discussed earlier, the alternation feature of cooperative education and the accompanying delays in graduation make advising even more critical.

Faculty must not only provide advising, but the geographer must assume the role of liaison with the nonacademic community to market geography majors to employers. Some universities may provide field study, internship, or cooperative education offices that can assist in these functions. Nonetheless, geographers can not rely totally on those offices to cultivate experiential opportunities for majors because their staffs may lack the knowledge of geographers' capabilities. Also, many of the benefits of interacting with nonacademics will be forgone if the contacts are brokered by a third party. On the other hand, the services of such offices should not be ignored; their expertise, support services, and existing employer networks will augment the productivity of a geography 'sales campaign.' One liaison will be sufficient in most departments, but that individual will need the assistance, ideas, and support of other faculty. Such support is particularly important because the liaison's functions, while important to the discipline, will generally be of little consequence in the reward systems of higher education.

Another personnel function will be the monitoring of progress of the co-op students on work assignments. Departments should assume supervisory roles with caution because supervision implies legal liability. Nonetheless, a faculty member should monitor the co-op students to evaluate the academic outcomes of their work and to assist students in integrating work with their educational programs. To accomplish this function, an experientially related course might be introduced. Historically, cooperative education institutions have offered orientation courses that cover career decision making, résumé development, interviewing, and job-search techniques. Because such information can be acquired on campus outside a geography department, an orientation course may be redundant. Another option would be to offer a post-work assignment course. This course could be a capstone course, as advocated by Morrill (1983), or it might incorporate the relevant placement information outlined by Rafferty (1977). A more realistic approach, given the scarcity of faculty resources, may be the monitoring of students' progress through a series of work reports and evaluations.

It should be evident, then, that making suggestions concerning specific curriculum modifications to enhance a departmental cooperative education program is difficult because faculty strengths will vary. Perhaps more important than the exact curriculum is a commitment by the department to cultivate linkages with the nonacademic world.

Summary

Applied geography has stimulated much debate over the past decade. One outgrowth of this dialogue has been a rising concern with the role of experiential education in geography. Geography's participation in cooperative education has been growing, and one can assume in the absence of data that the number of internship programs has likewise increased. Such trends are healthy for the discipline because experiential education adds a career development dimension to educational programs that cannot be duplicated in the classroom.

Experiential education, especially cooperative education, possesses many advantages for departments and their majors but one of the most important is the provision of employer-based training. Such training precludes the need for departments to revamp academic curricula constantly to adjust to the diverse currents of the job market. Instead, departments can focus on the establishment of a quality curriculum incorporating liberal arts and generic skills and rely on employer-training programs to provide the specialized skills required for students to enter specific labor markets.

Even though cooperative education possesses advantages not inherent in internships, a department should offer both options to meet students' needs. For example, employer bias and the length of their academic programs may cause internships to be more feasible for graduate students than cooperative education. Also, those undergraduate majors converting to geography late in their college career may not have the time to engage in a co-op assignment.

Internships and cooperative education can be implemented without radical reorientation of departments. Departments can utilize existing faculty strengths coupled with some additional expenditure of faculty time for advising, liaison, and monitoring duties. Personnel reallocations will be more feasible in most institutions than acquiring sufficient monies to hire applied geographers. Such additions, however, may be desirable when new or replacement faculty positions arise. In essence, each department will have to assess its own potential and needs before developing experiential education goals.

Certainly, experiential education is not a panacea for all problems challenging academic units, but it can alleviate some pressures associated with erosion of career opportunities, declining enrollments, and the like. Departments and the discipline must market themselves to be successful. Perhaps the paramount virtue of an applied geography program is that faculty must begin marketing the discipline, thereby discarding the insularity of higher education, and fostering interaction with nonacademic geographers.

References

Beard, D.P. 1976. Professional problems of nonacademic geographers. *Professional Geographer* 28:127–131.

Beaujeu-Garnier, J. 1975. The operative role of geographers. *International Social Science Journal* 27:275–287.

Bein, F.L. 1984. Extension geography: applied geography in action. *Professional Geographer* 36:236–238.

Carter, R.L., and Steinbrink, J.E. 1974. Geography: from the product concept to the marketing concept. *Professional Geographer* 26:1–7.

Cooperative Education Association. 1978. *A Directory of Cooperative Education*. Philadelphia: Drexel University.

Cooperative Education Association. 1986. *A Directory of Cooperative Education*. Philadelphia: Drexel University.

Dunbar, G.S. 1978. What *was* applied geography? *Professional Geographer* 30:238–239.

Ford, L.R. 1982. Beware of new geographies. *Professional Geographer* 34:131–135.

Foster, L.T. 1982. Applied geography internship: operational Canadian models. *Journal of Geography* 81:210–215.

Foster, L.T., and Jones, K.G. 1977. Applied geography: an educational alternative. *Professional Geographer* 29:300–304.

Frazier, J.W. 1978. On the emergence of an applied geography. *Professional Geographer* 30:233–237.

Frazier, J.W. ed. 1982. *Applied Geography, Selected Perspectives*. Englewood Cliffs: Prentice-Hall.

Gad, G. 1979. Internships in the liberal arts undergraduate programme. *Journal of Geography in Higher Education* 3:15–23.

Hall, P. 1983. Surprise! Liberal arts students make the best managers. *Business Week's Guide to Careers* 1:16–19.

Harrison, J.D. 1977. What *is* applied geography? *Professional Geographer* 29:297–300.

Harrison, J.D., and Larsen, R.D. 1977. Geography and planning: the need for an applied interface. *Professional Geographer* 29:139–147.

Hart, J.F. 1978. Geography and decision making. *Journal of Geography* 77:252–253.

Heiges, H.E. 1972. A student internship program in geography. *Journal of Geography* 71:458–467.

Heiges, H.E. 1977. Progress and development of a student internship program in geography. *Journal of Geography* 76:147–149.

Hornbeck, D. 1979. Applied geography, is it really needed? *Journal of Geography* 78:47–49.

Jumper, S.R. 1975. Going to the well. *Professional Geographer* 27:419–425.

Millonzi, J.C. 1979. Liberal education and cooperative education are they really compatible? *Journal of Cooperative Education* 15:5–13.

Monte, J.A. 1983. The job market for geographers in private industry in the Washington, D.C. area. *Professional Geographer* 35:90–94.

Morrill, R. 1983. The nature, unity, and value of geography. *Professional Geographer* 35:1–9.

Rafferty, M.D. 1977. The geography placement seminar: a course syllabus. *Professional Geographer* 29:215–217.

Richardson, D. 1986. An applied geography agenda. *Applied Geography Speciality Group Newsletter* 6:1–2.

Ritchie, R.J. 1983. Preparing liberal arts students for careers in business. *Journal of College Placement* 43:53–56.

Salter, C.L. 1976. The case for the nontechnician geographer in the technician's era. *Journal of Geography* 75:70–76.

Salter, C.L. 1983. What can I do with geography? *Professional Geographer* 35:266–273.

Sengenberger, D.L. 1975. The contribution of geography to general education. *Journal of Geography* 74:7.

Smith, B.W., and Spinelli, J.G. 1979. A development program for geography: planning in the present for the future. *Journal of Geography* 78:70–76.

Spinelli, J.G., and Smith, B.W. 1981. Cooperative education versus internship: a challenge for an applied geography programme. *Journal of Geography in Higher Education* 5:163–168.

Stutz, F. 1980. Applied geographic research for state and local government: problems and prospects. *Professional Geographer* 32:393–399.

Stutz, F., and Heiges, H.E. 1983. Developing courses in applied geography for transportation planners. *Professional Geographer* 35:206–210.

Bruce W. Smith
Department of Geography and
Cooperative Education Program
Bowling Green State University
Bowling Green, OH 43403
U.S.A.

5. Undergraduate Introductory Human Geography Textbooks: Less Theoretical and Basic, More Descriptive and Applied

Introduction

Geographic education is a subfield of geography that deals with the teaching of geographic concepts and techniques. Emphasis is placed on what to teach as well as how to teach it. Undergraduate geographic education has undergone many changes recently, reflecting geography's expanded scope of inquiry or research, and the adoption of new computer related technologies. As a result, undergraduate curricula and textbooks have changed to include such topics and techniques as locational analysis, quantitative methods, remote sensing, and computer carto-graphy. These new curricula and textbooks also reflect the respective authors' attitudes toward how geographic concepts and techniques should be taught, using either a 'basic' or an 'applied' research perspective.

Basic or pure geographic research seeks to expand the discipline's theoretical base and to answer the universal questions of *where* and *why*. These questions are basic to the science, which studies the location, spatial arrangement, and spatial interaction of both the physical and human environments, as well as the processes that generate those spatial distributions. Applied geographic research seeks to use geographic concepts and technique to solve a wide range of everyday problems: from the evaluation of toxic waste dumps to the delimitation of a market area for a proposed supermarket. This past decade has seen a shift in undergraduate curricula and textbooks to the applied research perspective, as geographers seek to show students and the general public that geography is relevant in today's world.

The purpose of this paper is to examine a collection of introductory human geography texts (IHGTs) published within the last fifteen years to document their shift toward applied geography, and to identify the reasons and implications of such a move. I selected IHGTs to study because they serve as a vehicle to introduce undergraduates to the fundamental concepts and techniques of human geography, and because they are the foundation upon which undergraduates build their knowledge of geography.

The textbooks selected for examination had to meet several criteria: they had to be human (not cultural, social, or economic) geography texts, they had to be written for lower-division students;[1] they had to be published within the last fifteen years, which would encompass the quantitative and post-quantitative revolution eras; and they had to be 'American' in origin – i.e., texts with an American author(s) or texts written by an author(s) who primarily taught in and wrote for the American university market.[2]

M. S. Kenzer (ed.), Applied Geography: Issues, Questions, and Concerns, 65–72.
© 1989 *Kluwer Academic Publishers.*

Appearance and Purpose of Human Geography Texts

Prior to the 1970s, most introductory texts were written for courses in 'Introduction to Geography,' 'Physical Geography,' or 'World Regional Geography.' While many texts contained a great deal of human geography material, there were none devoted solely to human geography's concepts, theories, or research trends. By the late 1980s, however, fifteen texts meeting the aforementioned criteria would be published: Cox 1972; Fielding 1974; Morrill 1974; De Blij 1977; Morrill and Dormitzer 1979; Larkin, Peters, and Exline 1981; de Blij 1982; Lowe and Pederson 1983; Rubenstein and Bacon 1983; Getis, Getis, and Fellmann 1985; Jackson 1985; de Blij and Muller 1986; Harries and Norris 1986; Stoddard, Blouet, and Wishart 1986; Austin, Honey, and Eagle 1987.

Generally, the texts of the 1970s were primarily concerned with presenting human geography from a theoretical/analytical approach, which emphasized a basic research perspective. Cox (1972, p. v), for example, wrote his textbook because he 'found it increasingly difficult to locate textual materials that adequately reflected' the theoretical/analytical revolution that had begun in the 1960s. Fielding (1974, p. xv) readily admitted that his text was 'biased toward a theoretical approach to the study of human behavior,' while Morrill (1974, p. xi) stated that he intended to 'examine and define the general principles that constitute a theory of spatial organization.'

Texts published in the 1980s, however, primarily focused on a descriptive approach to human geography and how such an approach could be used to solve world problems, thus introducing students to an applied research perspective. Larkin, Peters, and Exline (1981, p. iii), for example, hoped to 'provide a geographical framework for the analysis of current world problems,' while Rubenstein and Bacon (1983, p. xxv) emphasized 'the relevance of geographic concepts to human problems.' Austin, Honey, and Eagle (1987, p. xiv) continued this trend by stating that their text 'focuses most on those elements and characteristics of the world around us which are of greatest relevance for the world our students will experience.' The change to a descriptive approach with an applied research perspective was typical of texts written in the 1980s, with the sole exception of Lowe and Pederson (1983).

Defining Human Geography

While an examination of the fifteen IHGTs revealed an equal number of definitions of human geography, the definitions of the 1970s texts generally differed from those of the 1980s by presenting primarily theoretical/abstract definitions. Cox (1972, p. 3) defined human geography as being 'concerned with the description and analysis of locational patterns of static or moving phenomena of human origin on the surface of the earth,' while Fielding (1974, p. 5) stated that human geography was concerned with 'the locations and arrangements of phenomena on the surface of the earth and the processes that generate these distributions.'

The definitions of the 1980s, however, were descriptive/concrete. Rubenstein and Bacon (1983, p. xxv) defined human geography as the study of 'human characteristics, such as language, industries, and settlements,' while Jackson (1985, p. v) stated simply that human geography 'is the geography of human kind.' De Blij and Muller (1986, p. 3) similarly defined human geography as 'primarily interested in the human world, in human activities and behaviors, and the marks these make on the land.'

General Contents of Textbooks

What is (or is not) included in a text can imply what knowledge is (or is not) important to learn; a text's contents will also affect the type of foundation a student will have to build upon. The 1970s texts, which expressed the desire to emphasize geographic theory and current basic research trends, contain chapters dealing with 'Locating Nodes: Some Basic Concepts' (Cox 1972), 'Hypotheses and Research' (Fielding 1974), and 'The Central Place System' (Morrill and Dormitzer 1979).

In contrast, the 1980s texts contain chapters dealing with major subfields of geography. Rubenstein and Bacon (1983), for example, has chapters on 'Population' and 'Political Geography,' while Getis, Getis, and Fellmann (1985) has chapters focusing on 'Economic Geography' and 'Behavioral Geography.' Many 1980s textbooks also refer to various aspects of a subfield rather than to that subfield as a whole. Harries and Norris (1986) write about the 'Geography of Language' and the 'Geography of Religion,' while Jackson (1985) writes about 'Migration' and 'Settlement and Land Tenure.'

Finally, in keeping with their applied research perspective, several 1980s texts contain material dealing with careers in geography. The Getis, Getis, and Fellmann (1985) text, for example, contains a section entitled 'The Purposes of Geography: What Geographers Do,' while the Harries and Norris (1986) text contains a chapter on 'Geographers at Work.' Morrill and Dormitzer's (1979) 'Afterword: But What Do Geographers Do?' is an exception.

Comparison of Two Textbooks

To illustrate further the differences between the 1970s and the 1980s texts, the Fielding (1974) and Rubenstein and Bacon (1983) texts were selected for extended comparison. These particular texts were chosen because I have had a great deal of experience using both of them in the classroom; based on personal analysis, they respectively epitomize the two extremes of texts discussed in this paper.

The major difference in the two texts can be seen in how each introduces students to the fundamentals of human geography. Fielding's first three chapters, which represent 38 percent of the text, are devoted to geographic fundamentals and an introduction of geography as a social science. Chapter One, 'Geography and Human Behavior,' defines geography as an observational science. The scientific concepts of assumption and surrogate variables are discussed, as are the geographic concepts of agglomeration, distance, accessibility, interaction, utility, and satisfaction. Chapter Two, 'Analysis of Spatial Distributions,' presents the concepts of discrete, continuous, and contingent spatial distributions, random and nonrandom patterns, regions, sampling, sampling designs, and graphic space models. Finally, Chapter Three, 'Hypotheses and Research' discusses inductive and deductive research, hypothesis testing, and regression analysis. The research example presented in the section 'Approaches to Geographical Research' exemplifies the text's basic research perspective with its discussion of hypothesis development, independent and dependent variables, and regression analysis.

Rubenstein and Bacon introduce geographic fundamentals in one chapter, 'Basic Concepts,' by asking three questions. The 'What do geographers do?' question focuses on the historical development of geography, the human-land approach, and careers in geography. Students are told of career opportunities in academia, government, and private business. Thus, students are introduced to the applied aspects of human geography at the onset of an introductory course. The second question, 'What concepts are used to explain where people and activities are found?' introduces students to location, density, concentration, pattern, and map projections and scales. The last question, 'What concepts are used to explain why people and activities are arranged in a particular way?' centers on physical environment factors and uniform and functional regions. The three questions are answered in nontechnical terms using many concrete examples, rather than using abstract examples as was often the case in the Fielding text.

A final comparison of the two texts centers on student reaction to the texts. Generally, I found that the Rubenstein and Bacon text was greatly preferred to the Fielding text. Comments about Fielding's text included remarks like, 'When are we going to study geography?,' 'I should have taken a stat course before I took this class,' and 'What has this got to do with everyday life?' The Rubenstein and Bacon text provoked a completely different set of reactions. Students liked the text, citing ease of comprehension (which could have been a function of the writing style), its interesting topics, and its relevancy in understanding world problems. Students were also pleased to see mention of specific careers that one could pursue with a geographic training.

From Theoretical/Basic to Descriptive/Applied

A textbook's content, approach, and research orientation generally reflect *when* and *where* the author was trained. Geographers trained in departments influenced by the quantitative revolution of the 1960s are inclined to use quantitative methods, modelling, and theory building. These same geographers have a propensity to engage in basic research. Other geographers were influenced by the 'socially caring years' of the late 1960s and 1970s, rather than the quantitative revolution. They seem to have a propensity toward applied research.

The decades of the 1970s and 1980s brought forth still other forces that have influenced the type of texts that are published today. Among those forces are curriculum relevancy and careerism. During the 1970s there was a nationwide trend toward making college curricula relevant to the 'real world.' Students wondered how courses in eighteenth-century English literature or Medieval art were going to be of any use to them once they graduated. This theme of usefulness was carried over into the 1980s where students are now pursuing a college education with specific career goals in mind. Curricula that are career specific, such as accounting and business administration, have been very popular, while less career specific curricula, such as history or English literature, have not been as popular.

Given the themes of social awareness, curriculum relevancy, and careerism, many geography departments are emphasizing an applied curriculum rather than one based solely on the theoretical approach with its basic research perspective. This shift to an applied curriculum has created a need to publish textbooks that espouse this philosophy, hence the development of IHGTs texts with an applied research emphasis.

Implications of the Change to an Applied Emphasis

Positive results from the change to an applied emphasis in the undergraduate curricula and introductory texts are increased enrollments, an interest in majoring/minoring in geography, an interest in graduate education in either geography or urban planning, and the marketability of the undergraduate with a degree in geography. I began to notice both an increase in enrollments and an interest in geography after I began using a descriptive/applied text and once I began to emphasize applied geography in my 'Introduction to Human Geography' course. Students commented that they liked the career aspect of the course, and that they found the text to be quite interesting and useful. I moved away from a theoretical/basic research approach and text, and toward the applied approach, when I realized that we (geographers) were appealing to a very small audience with the approach we were taking. I feel that geography should broaden its base to produce professional geographers at all levels of education, not just at the doctoral level.

Enrollments in the 'Introduction to Human Geography' class have also increased because the University of South Alabama's Department of Marketing and Transportation decided to require their 200 majors to take the course after reviewing my course outline and the Rubenstein and Bacon text. The marketing faculty were impressed by the usefulness of the material and by how well it fit the career needs of their students.

The last three years have also seen an increase in the number of our majors going on to graduate school. The applied aspect of the University of South Alabama's curriculum has influenced many students to seek further training in the fields of computer cartography, geographic information systems, and urban planning. The success rate of these students has been high, indicating that their applied undergraduate education had prepared them sufficiently for graduate studies in applied departments such as the University of South Carolina. Still other students have gone straight into the job market from undergraduate school. Some of these students have entered the armed forces' intelligence branch, where they can use their map interpretation and remote sensing skills, while other students have been employed by travel agencies, where they can use their knowledge of local and foreign places.

While there are many positive impacts regarding the change in both curricula and texts to an applied perspective, there are some negative impacts as well. Many students may not receive the foundation they need for upper-division courses or for graduate programs that may use a theoretical approach and emphasize basic research. These problems could occur if a department has not fully adopted an applied curriculum, or if a student chooses a graduate program that emphasizes basic research. In either case, the student would have some difficulty handling the theoretical coursework because their foundation courses would not have emphasized the theoretical aspects of human geography.

Another area of concern centers around current and future career preparation. There is the distinct possibility that a department's applied curriculum may not accurately reflect current job skills because the department's curriculum may be emphasizing outdated methods and/or using obsolete equipment. This situation can easily occur if a faculty does not keep abreast of current changes and/or if a department cannot afford the latest in computer based technology. Students graduating from this type of department could find themselves ill prepared for the current job market. It is also possible that a student may not be given the requisite conceptual and research skills to cope with changes that might occur in the future. This is a real possibility, as much of an applied curriculum revolves around computer related technology; and given that this technology is dynamic, students may not be able to make the necessary adjustments needed to meet new job skill demands.

Summary and Conclusions

The number of IHGTs written since the 1960s reflects a need to have specific texts for the human branch of geography. Those of the 1970s primarily used a theoretical/analytical approach with an emphasis on basic research, while those texts of the 1980s primarily used a descriptive approach with an emphasis on applied geography. The difference in conceptual approach and research perspective has affected the purpose and content of these texts. Fielding (1972) and Rubenstein and Bacon (1983) were compared with special attention given to their respective section(s) on the fundamentals of geography. Their differences epitomize the differences between the two types of human geography textbooks.

The academic training of textbook authors, social awareness, curriculum relevancy, and careerism were identified as reasons why there was a change in curricula and in the type of IHGTs written in the 1980s. The changes helped to increase enrollments, interest in geography and graduate education, and the marketability of undergraduates. These gains, however, could be offset by students not learning appropriate marketable skills or by being ill prepared for graduate programs like that at The Ohio State University, which emphasizes the theoretical approach in such research areas as climatology and the diffusion of innovations.

Even though there are some negative ramifications, overall I strongly feel that geography departments should continue the movement toward an applied curriculum and related texts, not only at the baccalaureate level, but at the master's level as well. In doing so, geography departments will be expanding the marketability of their graduates beyond academia. Geographers need to broaden their job search, as universities have limited employment opportunities and require doctoral degrees. The profession of geography should be making every effort to increase the marketability of its students at all levels of education, as have the disciplines of geology and economics. The profession needs to move away from the rarified atmosphere of campus, and begin to engage actively in solving problems that face our everyday world. Geographers who do basic research also need to lessen their 'holier than thou' attitude, which has hindered the development of applied curricula and discounted the importance of applied research. These two changes must take place or geography will never take its rightful place in influencing public and private decision making. Geography needs to broaden its base and become part of everyday life. The profession can begin to accomplish this goal by expanding applied geography's role in undergraduate and master's level education.

Notes

1. Abler, Adams, and Gould (1971) was not included because it was not intended for the freshman-sophomore audience.
2. Bradford and Kent (1977), Chapman (1979), and Haggett (1972) were not included in the study because they did not meet the 'American' criterion. See Johnson and Claval (1984) for a discussion of the different geographical perspectives found around the world.

References

Abler, R., Adams, J.S., and Gould, P. 1971. *Spatial Organization: The Geographer's View of the World*. Englewood Cliffs, NJ: Prentice-Hall.

Austin, C.M., Honey, R., and Eagle, T.C. 1987. *Human Geography*. St. Paul, MN: West.

Bradford, M.G. and Kent, W.A. 1977. *Human Geography: Theories and Their Applications*. Oxford: Oxford University Press.

Chapman, K. 1979. *People, Pattern, and Process: An Introduction to Human Geography*. NY: Halsted.

Chorley, R.J. and Haggett, P., eds. 1968. *Socio-Economic Models in Geography*. London: Methuen.

Cox, K.R. 1972. *Man, Location, and Behavior: An Introduction to Human Geography*. NY: Wiley.

de Blij, H.J. 1977. *Human Geography: Culture, Society, and Space*. NY: Wiley.

de Blij, H.J., 1982. *Human Geography: Culture, Society, and Space.* 2d ed., NY: Wiley.

de Blij, H.J., and Muller, P.O. 1986. *Human Geography: Culture, Society, and Space.* 3d ed., NY: Wiley.

Fielding, G.J. 1974. *Geography as a Social Science.* NY: Harper and Row.

Getis, A., Getis, J. and Fellman, J. 1985. *Human Geography: Culture and Environment.* NY: Macmillan.

Haggett, P. 1972. *Geography: A Modern Synthesis.* NY: Harper and Row.

Harries, K.D. and Norris, R.E. 1986. *Human Geography: Culture, Interaction, and Economy.* Columbus, OH: Merrill.

Jackson, W.A.D. 1985. *The Shaping of Our World: A Human and Cultural Geography.* NY: Wiley.

Johnston, R.J. and Claval, P., eds. 1984. *Geography Since the Second World War: An International Survey.* Totowa, NJ: Barnes and Noble.

Larkin, R.P., Peters, G.L., and Exline, C.H. 1981. *People, Environment, and Place: An Introduction to Human Geography.* Columbus, OH: Merrill.

Lowe, J. and Pederson, E. 1983. *Human Geography: An Integrated Approach.* NY: Wiley.

Morrill, R.L. 1974. *The Spatial Organization of Society.* 2nd ed., North Scituate, MA: Duxbury.

Morrill, R.L., and Dormitzer, J.M. 1979. *The Spatial Order: An Introduction to Modern Geography.* North Scituate, MA: Duxbury.

Rubenstein, J.M. and Bacon, R.S. 1983. *The Cultural Landscape: An Introduction to Human Geography.* St. Paul, MN: West.

Stoddard, R.H., Blouet, B.W., and Wishart, D.J. 1986. *Human Geography: People, Places, and Cultures.* Englewood Cliffs, NJ: Prentice-Hall.

Victoria L. Rivizzigno
Department of Geology-Geography
University of South Alabama
Mobile, AL 36688
U.S.A.

SECTION III

Considerations in Physical Geography

6. A Climatologist's Personal Perspective on Applied Geography

Personal View and Some Professional Background

Climatology, part of geography, is the study of atmospheric conditions in relation to the earth's surface. Climate is a component of the earth-atmosphere system with all its constituent natural, human-influenced, and artificial ecosystems.

The viewpoint of a geography divided into two parts, one a natural science and the other belonging to the social and behavioral sciences and the humanities, is supported by many American and European geographers. It seems that this dualistic notion rests on a false dichotomy (James 1967, p. 21); geography is a holistic discipline, a position that has a history of common acceptance in the United Kingdom (Johnston 1985, p. 11) and a few years ago, at least, was gaining in the United States (Johnston 1983b, p. 3). To elaborate somewhat, I see geography as a conceptually unified earth science without teleological focus (Leighly 1983, pp. 86–87), which studies the earth-atmosphere system ('the environment') ecologically (Hare 1977, p. 264) and thus synthesizes knowledge of the living and nonliving contents of the earth's surface (Marcus 1982, p. 89; 1979). Except, perhaps, for its nonanthropocentric element, this holistic perspective and its consequent irreverence for subdisciplinary and even disciplinary boundaries is, as will become clear, valuable to much applied work (Buttimer 1979, pp. 11–12, 34).

Many global geography treatises do not reflect this point of view. They tend to be human-centered and, moreover, often are disjointed in that they discuss a number of themes, topics or 'traditions' in succession without making the linkages between them very evident (Beaujeu-Garnier 1983, pp. 150–51). To what extent philosophical reasons are responsible is not always clear, nor does acceptance of the holistic approach guarantee an integrated discussion; but, a model I should like to hold up is Haggett's synthesis (Haggett 1983). The viewpoint expressed here does not preclude the development of concentrations (subfields) such as urban climatology, themes, such as mankind's place among other living communities, or the characterization of place or space.[1] Individually, they may well be human-centered, but their pursuit in isolation makes as little sense as the study of pediatrics or neurology would without constant reference to the entire field of medicine. Connections with the rest of geography should be kept in mind and analyzed to the degree necessary to put the focus into clear (historical) perspective.[2] As was recognized by Stoddart two decades ago (Gregory 1978, p. 47), the adoption of the principles of systems theory for geography at all scales – work that stands to gain from the ability to combine 'the points of view of the mammoth and the microbe' (Hägerstrand 1983, pp. 248, 254) – is particularly advantageous to the approach of oneness and connectivity championed here and perhaps, also on account of the

M. S. Kenzer (ed.), Applied Geography: Issues, Questions, and Concerns, 75–98.
© 1989 Kluwer Academic Publishers.

theory's alleged greater explanatory and predictive powers, to the addressing of societal problems (Johnston 1983b, pp. 41–43; Beaujeu-Garnier 1983, pp. 150–51).

For the last dozen years or so, my emphasis has been on climatology, whereas during the previous six or seven years, following the completion of my dissertation (Lier 1968), I concentrated on the human aspects of geography. It is of some comfort to discover that in this shift I share at least one career attribute with such geographers as Stephen B. Jones (Harris 1985), Leighly (Parsons and Vonnegut 1983, pp. 3–7), Nieuwolt (1977, book jacket), and Thornthwaite (Parsons and Vonnegut 1983, pp. 8–10). Largely by example, I shall discuss my view of applied geography, consider the applied nature of climatology, and try to come to a conclusion about applied geography's value to the discipline.

Applied Geography

Logic demands that 'applied geography' be used for geographic knowledge that has been applied to something (other knowledge, education, practical problem solving), but practice in geography and in other fields has sanctioned 'applied' for uses where 'applicable' would have been grammatically correct. In this paper, rather than courageously but lonely striking out across the hummocky fields, I join those who have already strayed onto the by now well-worn trail and hereby offer the following as a working definition of applied geography: the application of geographic knowledge to other knowledge in formal education or to practical problem solving outside academia for the real or perceived benefit of society or the environment. Hence, much of the history and prognosis of geography could be written in terms of applied geography.

Internal and External Applied Geography

Applied geography, as defined here, has two elements, an internal and an external one. The former concerns the academic and the larger educational communities. This includes geography courses (or other courses administered by geography departments) that perform a service function for other parts of the university. It also includes courses or programs specifically designed to enhance the employability of students as geographers outside academia (some cartography and planning courses, courses in environmental impact report writing, and so on). The geographic alliances now spreading across the U.S. in a welcome effort to raise the low level of geography teaching in the country's elementary and secondary schools are another example.[3] The concentration of this paper, however, will be on the second, external type of applied geography, which I shall from now on simply refer to as applied geography. All of it is fieldwork, broadly defined, in that it is dialectically related to our lives on campus (Pred 1983, pp. 91–92), a quality shared with many (possibly most) off-campus experiences, from foreign travel to the mundane.[4] The intensity of these dialectical connections will vary with one's personality, age, the position on one's life path, the poignancy of path convergences, and other experiences (Pred 1983).

Habermas (1973, p. 8) distinguishes in any science between (1) the logical structure and (2) 'the pragmatic structure of the possible applications of the information generated within its framework' with 'a systematic relationship' between the two. The magnitude of the latent applicability of the pragmatic structure varies from small in the 'pure' to large in the 'applied' sciences (or divisions of sciences), and depends, in part, on the influence of society on science. This means that praxis affects theory, which is part of a reciprocal relationship. Reminiscent of the latent heat of vaporization that can become manifest, this applicability can too. When it does manifest itself, it is available to society as a tool for the creation of benefit, that is as theory that affects praxis and thus establishes the reciprocity (Habermas 1973, pp. 253–54),[5] like the dialectical relationship discussed earlier. The frequency with which this happens is a function of the exposure of the science to society and no science develops in total isolation (Kuhn 1977, pp. xv, 32n; Wright 1966, pp. 62–64). Since geography's exposure is particularly great (Beaujeu-Garnier 1983, pp. 141–43; Lowenthal 1961, pp. 241–42), 'pure' geography – in quasi-separation from applied geography (Porter 1977, pp. 282–83) – exists somewhere near the least pure end of the spectrum.

Background

Phlipponneau surveyed the development of applied geography (emphasizing the twentieth century), pointing out that although applied geography had earlier beginnings, the great push came after World War II (Phlipponneau 1981, pp. 134–35). Of major importance was the establishment in 1976 of the Working Group on Applied Aspects of Geography by the International Geographical Union (I.G.U.). Its aims were to 'delimit the outlines and content of applied geography ... study problems relating to the further training of young geographers who plan to pursue a career in applied geography [and] ... list employment opportunities for the young geographers so trained' (Phlipponneau 1981, p. 135). Although applied geography was not defined, the statement clearly focused on off-campus employment. Phlipponneau indicated the strong interest in the practical application of geography in the U.S.S.R.,[6] other socialist countries, and in the developing world (1981, pp. 150–53), and by alluding to the significance to applied geography of Renaissance exploration and of the inventory taking in newly conquered territories (p. 133), he indicated that the application of geographic knowledge has never been the exclusive domain of formally recognized geographers; it predates academic geography by a few centuries. We can, in fact, go back much further and confidently assume that the principle of applied geography is as old as mankind.

Envision paleolithic people who, while moving about and prompted by curiosity, material need, and exposure to their world, gained geographic knowledge for the benefit of food quest and other purposes. Some of this knowledge was applied soon after it was acquired, some later, but probably all of it affected their spatial behavior and other environmental relations. This gave rise to further insights and a time-sequence of geographic (and other) knowledge and application was created. Some observations may have involved differences in vegetation densities, which, after lengthy sojourns, may have led to the detection of relationships with moisture supply. Similarly, members on the home front may have, in time, identified spatial and seasonal differences in recurring weather and climate that influenced spatial decisions, as with the siting or resiting of encampments. Observations of territory beyond the immediate vicinity of the home site would have permitted comparisons that induced or precluded movement into new country, and so on. We know that there were people among those of traditional societies who applied their geographic knowledge to map making for their own purposes, benefitting many travelers from elsewhere as well. Examples come from the Eskimo of the eastern Canadian Arctic (Spink and Moody 1972), Alaska, from Kropotkin's travels in Transbaikalia, from North American Indians, and from many other locales (Adler 1911).[7]

Application and the Growth of Knowledge

Premodern peoples were and are active applied geographers with dialectical links with their environment or, put another way, with reciprocally reflexive relation-ships to their environment; the acquisition of knowledge, its practical applications to societal problems, and the acquisition of subsequent knowledge during the applications developed in a *continuous* self-reinforcing process (Buttimer 1984, p. 21). Some of what they learned may have found application only after a long time and some none at all. Although the application of Habermas's dual structure seems rather excessive in this instance, part of the premodern lived experience was related to utility with praxis serving both as the source and the touchstone of ideas about their world. As Goethe observed, '[*die Praxis*] ist der Prüfstein des vom Geist Empfangenen ...' (Dobel 1968, Column 726).

Many documented examples of both premodern and modern peoples can be found of pertinent, lay geographic knowledge and its practical applications. Native fishermen near today's Jakarta have, for centuries, made use of one of climate's true periodicities to support their way of life: the directions and reliability of local winds. In their small boats (designed to sail only before the wind) they set out at night on the landbreeze and, toward noon, leave their fishing grounds for port on the seabreeze (Braak 1925?, p. 85). Military history is replete with strategies heavily influenced by practical application of climatic knowledge. Modern examples include the timing of the two recent invasions of Russia[8] and the landing in Normandy. Another illustration comes from a much published boatman who told the story of how a lockkeeper in West Germany's Lahn Valley used his knowledge of the connections between weather, stream flow, and tree phenology to predict a drop in water level following the first signs of green in the spring (Pilkington 1970, p. 137). Many lay people practice geography and applied geography without knowing it (Beaujeu-Garnier 1983, p. 149), and it is not difficult to find other examples from before or after the rise of academic geography. In all cases the fund of geographic knowledge grows by using praxis as the touchstone of ideas in a process of two-way reflexivity.

The Problem of Definition

In 1981 the Association of American Geographers (A.A.G.) published a *Directory of Applied Geographers*, which was 'an attempt to provide a useful guide to the broad range of expertise available among geographers who work for a variety of business, consulting, or government firms of agencies' or who are wholly or partly self-employed (Anonymous 1981b). Off-campus employment as geographer was thus the selective criterion for inclusion, the same restrictive criterion seemingly used by the I.G.U. Working Group. Table 1 compares all geographers listed with those in climatology and classifies the latter's employment. Almost all geographers, including all climatologists, stated interests in more than one aspect of geography, usually in several aspects. Of the climatologists working for the federal government, four were employed by the U.S. Geological Survey, while smaller numbers were dispersed elsewhere. As to the climatologists' job descriptions, twenty-four referred to or strongly suggested geography. In order of frequency they were atmospheric scientist (6), planner (5), cartographer/photo interpreter (4), environmental scientist/environmental impact analyst (3), geographer (3), remote sensing specialist (2), and soil scientist (1). Judging from its contents, the 'Applied Geography' section of the *Professional Geographer*, which dates from late 1982, also follows the definition of the *Directory*.[9]

Table 1. Geographers Listed in *Directory of Applied Geographers* (1981).

		Total	Percent
All geographers		1198	100.0
Climatologists employed by:			
U.S. government	14		
State governments	5		
Other govt. levels in U.S.	3		
Private firms	10		
Self-employed/retired	3		
	—		
		35	2.9

Source: Anonymous 1981b.

Now consider Table 2, which has been extracted from the two latest A.A.G. membership directories (Anonymous 1982, 1987). As of mid-1982 and mid-1987, it lists geographers with on- and off-campus employment who identified with applied geography – by indicating proficiency in applied geography, membership in the Specialty Group on Applied Geography, or both – and compares them with the entire A.A.G. membership.[10] The data show that employed applied geographers are a rather small and stable group (with about 7 percent of the A.A.G.'s membership in both years), that there are more applied geographers with off- than on-campus employment, and that the ratio between the two subgroups (3:2) remained unchanged over the five-year period. It seems safe to assume that the members of the larger subgroup define applied geography in the manner of the *Directory of Applied Geographers*: immediate solutions to problems of personal sustenance by means of employment outside academia. But many applied geographers with on-campus employment follow a different definition, as does the British, wide-ranging, resource oriented periodical *Applied Geography: an International Journal*, launched in 1980. Furthermore, many A.A.G. members concentrate on subfields that have a strong applied element, without necessarily indicating proficiency in applied geography or membership in its specialty group. How do they define applied geography? A case in point is the category of Geographic Information Systems (G.I.S.), introduced in the 1987 directory. A 10 percent random sample from that publication revealed that more than three times as many geographers with proficiency in G.I.S. and/or membership in the G.I.S. specialty group did not identify with applied geography in the above manner than those who did.[11] By mid-1987, the A.A.G. membership included 367 employed G.I.S. geographers (6.7 percent), of whom 198 (3.6 percent) were employed off- and 169 on-campus, a ratio of 5:6. By comparison, the 1987 directory listed 225 employed climatologists (4.1 percent), of whom only forty-eight had off-campus jobs (cf. Table 1) and 177 (3.2 percent) worked on-campus.

Table 2. Members of the Association of American Geographers: Employed Applied Geographers (EAG) by Job Location and All Geographers

	1982 Members	Percent	1987 Members	Percent
All geographers	5324	100.0	5479	100.0
EAG – on campus	148	2.8	153	2.8
EAG – off campus	222	4.2	235	4.3

Sources: Anonymous 1982, 1987.

How an individual geographer distinguishes between applied and pure aspects of geography, if at all, depends to some extent on his or her view of geography. It would seem that geographers who take the anthropocentric view – geography as the study of the 'home of man' (Sack 1980, p. 197) or the 'home of humankind' (Johnston 1985, pp. 6, 13) – are in a good position to detect a duality. To them, the applied notion (as directed to human benefit) is woven into the basic geographic fabric. To those who follow the view supported here, however, the question of geography as a whole seems irrelevant for lack of a central focus within the earth-atmosphere system. This, of course, is not true of some specific kinds of work, such as resource studies.

Applied and Pure Geography

I have argued that geographic knowledge is being applied continuously in all walks of life. Yet, although reciprocal reflexivity is a corollary, it may not be a *conscious* reflexivity. Taking a leaf from the natural sciences (Kuhn 1970, p. 42), it is awareness of the potential of reflexivity between pure and applied geography (again, theory and praxis) that is the most promising benefit to the entire discipline. The botanist's practice of reciprocal cross-pollination would be analogous to the reflexivity discussed here: a conscious process, artificial, but not inherently continuous. Good illustrations come from the interaction between development theory and work by geographers and others in the Third World (Dickenson *et al.* 1983; Reitsma and Kleinpenning 1985) (see below). To return to the *Directory* and Table 1, it follows that conscious reciprocal reflexivity between academic and nonacademic geography will grow with the relations between the off-campus groups and academia. One way to encourage such growth is via student internships in the off-campus locations, with a requirement that interns report their experiences to their departments during formal discussion sessions. Ideas from those sessions can be transmitted to the outside and fruitful developments can be generated, not only to benefit applied employment, but also research and teaching.

If we equate paleolithic people within the encampment with our academic departments and their movements on the food quest with our linked outside world, the relations between the two sides (before and after the institutionalization of geography) show greater differences of scale than of principle, a point to which I shall return. The application of geographic knowledge to achieve the immediate solution of individual or societal problems was practiced then as it is today, and the modern applied movement, particularly in its concern for the employability of geography graduates, still focuses on that aspect of immediacy. It has the advantage that the need for quick solutions is readily perceived and appreciated by many, both on- and off-campus, and therefore it meets little resistance. It is particularly this kind of applied geography that has been urged upon departments ever since the latest clamour for 'social relevance' in the late 1960s.[12] It has been adopted by some and advanced by anxiety over declining student enrollments, unsympathetic or disinterested university administrators, and by examples of failing departments (Johnston 1983a, pp. 207–08). It is intended to produce quick material results and has a high probability to do so.

Two Categories of External Applied Geography

Within external applied geography, two categories can be distinguished.

Category A. This concerns geographic research aimed at solving practical individual or societal problems in the short term. This is done by geographers who work alone or in small, say, professor-student, teams, motivated by curiosity, pedagogical needs, and project rewards. Multidisciplinary participation is not common. Examples include contract research by faculty (possibly with student assistance) to solve local problems, and geographic work by the geographers of Table 1.

As to the latter example, Roy Shlemon, a geographer listed in the *Directory* as the president of his own firm, has given an enthusiastic account of his work (which excludes climatology) and of the opportunities for applied physical geographers in his genre. 'Unfortunately,' he observed, 'many physical geographers are unaware of the potential for fun and fortune in their chosen field' (Shlemon 1979, p. 9). He pointed out, however, that the resulting technical reports – in the case of his firm some eighty of a geological and geomorphological nature over about ten years (Parsons and Vonnegut 1983, p. 93) – often neither produce new concepts nor are well documented. Furthermore, they are usually soon made almost inaccessible, and their value to subsequent research is, therefore, quite limited (Shlemon 1979, p. 10).[13] This procedure approximates what Wright saw as an impediment to scientific growth: keeping knowledge locked up (Wright 1966, p. 63). Shlemon's account suggests that this fun-and-fortune route, rather than benefit the discipline, is designed to benefit a client and thus build a reputation and financial security for the investigator who runs a business. As such, he or she is driven primarily by profit motive instead of by intellectual curiosity or a wish to advance the academic pool of knowledge. Direct benefit to society will, of course, depend on the nature of the project, but reflexivity with academia is limited. This could be enhanced by the employment of student interns and by the geographer-entrepreneur's use of the customary communication channels open to all professional geographers. Shlemon has used the latter freely with many publications to his name (Parsons and Vonnegut 1983, pp. 92–93). I do not know how representative he is of the geographers in Table 1, but I suspect that, on the whole, the fun-and-fortune type of applied geography benefits the profession only modestly at best, a combination of restricted dialectical interaction with academia and the profit motive.[14]

The problem of the profession being shortchanged is not unique to this kind of applied geography, however. To judge from Hägerstrand's experience (1983, pp. 252–53), many a geographer absorbed by the outside world, while continuing to do geographic work, is 'lost' to academic geography. It seems quite clear that intermittent or continuous applied research engaged in by faculty members, preferably with student participation, carries a much greater promise of advancing the profession intellectually, of taking 'living experience ... back home to the think tank' (Hägerstrand 1974, p. 51), and of taking think tank resources back to the outside, than is generally true of the type of work discussed by Shlemon[15] and the 'lost' geographers.

Category B. This category differs from the previous one in that (1) team work is more common, (2) both spatial and time scales are usually larger, and (3) multi-disciplinary participation is the rule. Research institutes are prominent here, but they are a mixed blessing. While they offer opportunities to both natural and other scientists – among whom geographers can and do provide a unique perspective (Stoddart 1986, pp. 57–58) – one is restricted in terms of problem selection but more so in terms of time than outside the institutional context (Claval 1984, p. 284). And, there is little chance to allow ideas to grow (Hägerstrand 1974, p. 51).

Among current international programs, there are those on polar research who focus on resources and are supported by the U.S. government. Another example is the Global Geosciences Program (started in 1987 with the support of the National Science Foundation (NSF)), attempting to use a systems approach to study processes affecting the earth's human habitability (Brewster 1987b). Despite its professional attractiveness to applied climatologists and other geographers and the prospect of it becoming even more so,[16] no geographers are participating in it at present, an absence blamed on their 'lack of lobbying' (Brewster 1987a). Karan wrote recently that, for Third World projects, the U.N. and other international agencies do not consider geographers the most suitable participants. These agencies treat Third World development planning 'as a variant of macro-economic management' and have 'the acceptance of rapid growth as the principal objective.' He implied that this has resulted in project domination by economists and specialists in public finance (Karan 1985, p. 470). The Enlightenment's vision of progress – only now gradually being questioned (Vale 1986) – and of science as both its 'source and exemplar' (Kuhn 1977, p. 106), seems to be embedded in the U.N. charter (Karan 1985, p. 472). On the other hand, Brookfield (1983, p. 36) reported that because of the heavy participation by geographers in UNESCO's program on Man and the Biosphere (MAB),[17] established in 1970, it has become known as the 'International Geographical Programme,' and he noted that geographers have been making their mark in Third World development studies since well before the latest advocacy of 'social relevance' in the U.S. (1983, pp. 33–34).

Great differences in the origin and nature of underdevelopment between nations create a need for different development strategies, a holistic approach, and still greater involvement. In the words of two recent authors, 'geographers should make a concerted effort to develop the study of spatial differentiation of development, material well-being, and social injustice into an applied science' (Reitsma and Kleinpenning 1985, pp. 396–97). But, any field participation from the developed world must expect the effects of a growing trend in Third World nations to engage their own scientists (Phlipponneau 1981, pp. 154–57). In this section, applied climatology has been considered mostly by implication. Let us now discuss it specifically.

Applied Climatology

Few of us are familiar with the details of the original climatologic fieldwork by Alexander von Humboldt, so usefully summarized by Stoddart (1986, p. 37). The nineteenth century also saw the growth of applied climatology at sea with the enormous expansion of ocean navigation and its evolution from wind to steam power. The British Admiralty mounted a series of expeditions, including the famous voyage of H.M.S. *Beagle* (1831–1836) under Captain Robert FitzRoy, to collect meteorological and many other kinds of data.[18] After facing much skepticism (from U.S. Navy captains and others) of the practical value of methodical hydrographical and meteorological data gathering at sea, Maury won his case for applied climatology when (in the 1840s and 1850s) the newly acquired knowledge resulted in the substantial reduction of sailing times, cost, and greater navigational safety (Williams 1963, pp. 150–51, 190, 192). Van der Stok has presented and discussed climatic and other data derived from seventy-seven years of observation in the waters surrounding the islands that now constitute Indonesia (Van der Stok 1897).

Probably the best known American applied climatologist was C. Warren Thornthwaite. Even though he wrote his dissertation on urban geography, his interests in climatology grew soon after its completion (1930) and, except briefly around 1934 – his last experience as a faculty member – it received his close attention for the next three decades (see Parsons and Vonnegut, pp. 8–10; Leighly 1964). An early advocate of systems study, he applied his knowledge to agriculture (irrigation in particular), industry, and waste disposal, while his research foreshadowed modern work on energy and moisture budgets. The establishment of the Laboratory of Climatology (New Jersey) was his crowning achievement. It has attracted students from all over the world and since 1948 has published the monograph series *Publications in Climatology*. After Thornthwaite's death in 1963 the laboratory was continued as part of C.W. Thornthwaite Associates. He never succeeded in linking his laboratory institutionally to academia (Leighly 1964; Marcus 1982, pp. 80–81), though there is at least one connection today: since 1979 (vol. 32, no. 1), *Publications in Climatology* has been issued jointly with the University of Delaware Center for Climatic Research.

Meteorology and Climatology

Both meteorology and climatology distinguish between basic and applied aspects, but the distinction is made more easily for the two sciences together than separately (a fact that may have led the American Meteorological Society to change its *Journal of Applied Meteorology* to the *Journal of Climate and Applied Meteorology* with the January, 1983 issue). Avalanche studies provide a good example of this interdependence and of the significance of other aspects of geography. Slope orientation, elevation, the weather three to five days before an avalanche, snow conditions, macro- and micro-relief, vegetation or the lack of it, and a host of other natural and anthropogenic factors have a bearing on the distribution of avalanches, their destructive force, potential of occurrence, and effect on the landscape (Salomé 1979).[19]

Another illustration of the interplay comes from deliberate meso- and even macro-scale weather and climate modification. Although research is often clearly meteorological, as when aimed at rain making, hail control, hurricane modification, and fog dispersal, in other cases climatologists are involved to greater or lesser degrees with studies on the removal of Arctic sea ice, large-scale water diversion, albedo control, and acid precipitation (Goudie 1986, pp. 261, 273–75, 277–81;

Oliver 1977, pp. 119–213; Jäger 1983, pp. 212–13). In applied meteorology the main emphasis is on 'nowcasting,'[20] and on short- and medium-range weather forecasting, all of which find applications across a wide spectrum of human activities and concerns. Climatology joins meteorology in the development of longer-term predictive capabilities and in the forecasting of extreme events (Jäger 1983, pp. 209–11) and in work on teleconnections (see below).

Roles of Climatology

That conditions of the lower atmosphere are monitored regularly in certain parts of the world for their proclivity toward air pollution is because climatology has identified the problem regions. It has also marked out regions particularly prone to acid precipitation, but much work remains, and it begins to look as if few parts of the world are immune. Who would have suspected Tibet to have a pH reading of 2.25 (Harte 1983)? Climatology can assist in estimating the extent of areas likely to be affected by nuclear accidents. Meteorologists must act at the time of disaster, but can be effective only to the extent that information on magnitude and height of atmospheric injection is supplied at once.[21] Climatology tells us which parts of the world have the greatest frequency of tropical cyclones or tornadoes or other severe weather. It is no accident that Miami (Coral Gables), Florida, and Norman, Oklahoma are the locations, respectively, of the National Hurricane Center and the National Severe Storms Laboratory. Environmental hazard research is the theme that unites these examples, drawing upon the expertise of climatologists and other workers (White 1973; Burton et al. 1978; Whyte and Burton 1980; Riebsame 1985, pp. 74–76). Further examples abound of the close association between the two applied fields. The pinpointing of D-day was largely a matter for the meteorologists, but planning it for late spring was based to a great extent on the climatological knowledge that conditions for crossing the English Channel by sea were likely to be more favorable than in late fall or winter and on the expectation of several months of relatively good campaign weather after a successful landing.

Teleconnections and Ecosystems

Research on teleconnections involves both climatology and meteorology and is important to long-range forecasting. It includes work on El Niño-Southern Oscillation (ENSO) phenomena (Barry and Perry 1973, pp. 379–417). The remarkable progress with ENSO research brings the day much nearer for predictions of anomalous events along coasts of the eastern Pacific and the corresponding lessening of human problems and ecological disturbances (Hare 1985, pp. 49–51; Kawasaki 1985, pp. 134–36; Wooster and Fluharty 1985). Teleconnection work began more than sixty years ago (Barry and Perry 1973, p. 414) and has given rise to an extensive literature that is constantly being supplemented by articles in the *Monthly Weather Review, Science, Nature* (London), the *Journal of Climatology,* and the *Quarterly Journal of the Royal Meteorological Society*.

Climatology enters into biological ecosystem studies, as in the ongoing work by Walter and Breckle (1985–   ). The climate diagrams, well known from the world atlas of the 1960s (Walter and Lieth 1960–1967), are also used in these later publications. Stoddart (1986, pp. 256–57) has cited examples of this type of work and warned that the complexity of some systems makes prior experience with simpler ones advisable. Since a great deal remains to be learned about ecosystems – a global survey has been cited as a particularly urgent task (Briggs 1981, p. 4) – the recent study of ecoregions in the U.S. by a geographer at the U.S. Environmental Protection Agency seems a step in the right direction (Omernik 1987).

Processes and Applications at the Earth-Atmosphere Interface

Applications of knowledge of present (and past) climates, especially of energy and water budgets help us to exploit snow as a resource in the northern Great Plains (Staple et al. 1960; Oke 1978, pp. 68–80); Willis and Carlson 1962; Sawatzky 1987) and to understand the nature and distribution of permafrost, which in turn has implications for engineering and agriculture (Mather 1974, p. 338; Denisov and Elovskaia 1979; Thompson 1974; Harris 1986). Soviet scientists are known for modern energy and water balance (budget) research; Budyko's atlas is one product of these concerns (Budyko 1963).[22] Questions continue to be raised about global climatic change and regional anomalies, their implications for the world's food problem, and about how to improve the application of climatic knowledge to decision making (Hare 1979). Much has already been written on food-climate-human relationships, as exemplified by the multidisciplinary volumes by Biswas and Biswas (1979) and Bach et al. (1981).

Small-scale spatial variety is intrinsic to alpine country and includes the constantly changing juxtaposition of sun and shade, offering a wealth of opportunity to the climatologist to study the heat and moisture budgets of different ecosystems at diurnal and seasonal scales (Miller 1981, 1977). They involve fluxes at the earth-atmosphere interface with which the promising concept of climatonomy is closely associated (Hare 1973, p. 189; Lettau, H. and K. 1969; Lettau 1969). The local application of this knowledge concerns the siting of crops, settlements, and other types of land use (Mather 1974, p. 303; Price 1981, pp. 419–41; Garnett 1935, pp. 601–17). Garnett saw in many of the surface patterns 'adjustments ... to critical factors of isolation ... brought about more or less unconsciously as a result of centuries of trial and error' (1935, p. 615).

Climatology and its applications hold a key to the solution of Sahel-type problems and, more generally, to an understanding of arid and semi-arid environments and desertification (Byrne et al. 1982; Berkofsky et al. 1981; Berkofsky and Wurtele 1986; Richter and Schmiedecken 1985; Gallais 1984; Watts 1983). The *Journal of Arid Environments* and *Applied Geography* are but two of several periodicals that frequently publish pertinent articles. Important also is the *World Map of Desertification* (1:25,000,000), a result of the U.N. Conference on Desertification of 1977 (FAO, UNESCO, WMO 1979?), and a bibliography of more than 1,500 entries, emphasizing Africa (Leng 1982).

An impressive example of large-scale climatological application to crop production comes from the Soviet Union, a nation with a particularly unfavorable agroclimatic base in most of its territory. During the Virgin Lands Program, initiated in 1954, millions of hectares of land in Soviet Asia were broken and planted primarily in wheat. Since they were known to be about one-half the length of an upper atmospheric standing wave removed from the traditional wheat areas of European Russia, which suggests a precipitation seesaw between the two areas, one aim of the project was to achieve the possibility of a good harvest in one area that would offset a crop failure in the other. In general, practice has borne out the theory (Lydolph 1979, pp. 222–23).

In the 1960s I found among the spring wheat farmers of Saskatchewan a preference for investing heavily in mechanized equipment that stands idle during most of the year to cooperative or other user arrangements. It was in large part a reaction to experience with the vagaries of the local climate under conditions of a short growing season. For most efficient operation, the implements must be available not only when the weather is dry and the crop ripe, but when its moisture content is low enough for combine harvesting (Lier 1971).

Agricultural geography and modern agricultural practices are, as Thornthwaite demonstrated, full of opportunities to apply weather and climate knowledge at all scales.[23] A few California examples come to mind: the work on viticulture by Holtgrieve and Trevors (1978) and Peters (1984, 1987), the raisin industry by Colby (1924) and Granger (1980a) and on other crops by Granger (1980b); but, it is not possible to do justice here to this important field of research.

Bioclimatology

Bioclimatology, like other crossover sciences, is an applied science by definition. Together with biometeorology it enters the realms of plant pathology and entomology and, among its concerns, studies relationships between atmospheric conditions and the development and spread of living organisms (spores, pollen, viruses). It studies links between weather and insect infestations in agriculture and forestry, and insect migrations, including the desert locust (Barry and Perry 1973, pp. 427–37). With other aspects of microclimatology it aids geneticists to develop varieties of crops that mature faster or are more resistant to drought, cool temperatures, frost, or other environmental parameters; it enters the decision making of individual farmers, and bears upon irrigation and rain-fed agriculture in numerous ways (Parsons 1986, pp. 378–82). Certain agricultural practices have long histories and among traditional ways of rural life today we can find methods that reflect the application of climatic knowledge dating from long before geography became a formal discipline. Examples include swidden farming and irrigated rice cultivation – beautifully contrasted and discussed as 'geographic' ecosystems by Clifford Geertz (1963) – Alpine and Mediterranean transhumance (Davies 1941; Evans

1940), and those given by Wilken (1972).[24] Research findings are often published in *Agricultural Meteorology, Agriculture and Environment, Agro-Ecosystems,* and *Archive for Meteorology, Geophysics and Bioclimatology.*

The Landscape

The landscape reflects the application of climatology in many ways. One thinks of the hedgerows of the Rhône Valley (Gade 1978), the green circles of central pivot irrigation in eastern Washington, the mountain country mentioned above, and the imprint of recreation and tourism. Though landscape paintings may tell us more about the artist than the scene and his or her knowledge of climate, some famous canvasses, as by Constable, Van Ruisdael, Turner, and Vermeer, not only appeal to the art lover, but also to the climatologist who may be perfectly insensitive to art.[25]

Flying west across California into the San Francisco Bay Area in spring, I notice the human-modified ecosystem of the snowy Sierra Nevada, the streams, artificial lakes, ditches, and green fields of the Central Valley, the wind mills of Altamont Pass, the reservoirs on the edge of the metropolis, and the salt evaporation ponds on the fringes of the bay. Yet, what would I have 'seen' without knowledge of the summer-dry climate; what if I had not been a geographer?[26] It is this 'duality of vision,' to use Stoddart's (1986, pp. 222–24) term that conditions the observation of facts, the recognition of process, and the formulation of ideas.

Modern urban areas present evidence of both successful and poor location with respect to climate. San Francisco International Airport is well sited in terms of the summertime advection fog that more frequently affects parts of the nearby region, but a roofless stadium only a few miles to the north has been plagued by high and chilly winds for some thirty years. There is insufficient coordination between urban planners and atmospheric scientists (Hobbs 1980, pp. 136–37). Landsberg and Oke have illustrated this with the location of industrial zones on the fringes of many eastern U.S. cities and with other aspects of urbanism (Landsberg 1981, pp. 255–61; Oke 1976, pp. 151–75). Applied climatology is important to housing and building design (Oliver 1979), the development of solar energy use (Anonymous 1981c), and to many other problems.

Perception

From a fresh perspective on the U.S., a Dutch geographer has shown that the concentration of federal installations and commercial enterprises in the Sunbelt and the selective migration of people from the American metropolitan heartland (the Snowbelt) to the southwest and south (the Sunbelt) have much to do with applied climatology. But, perception, aided by the newspaper press (which coined the Sun/Snow terms) must also play a significant role in migration and tourism, as it does in other situations (Eysberg 1984, pp. 123–35).[27]

Samuel Van Valkenburg used to tell students that among the officialdom of

Washington during World War II, a common notion was of Japan as a tropical archipelago with waving palm trees and other advertised South Seas trimmings. Many Europeans, Canadians, and even people in the eastern U.S. picture a California outside Los Angeles with blue skies and pristine air. Some are aware of air pollution in the Bay Area, but who imagines an 'early morning veil of brown smog low on the Fresno horizon with increasing frequency' or high ozone counts 'suspected of reducing cotton yields and of damaging the lower Sierra forests' (Parsons 1986, p. 386)?

The travel industry depends on actual climatic knowledge, but perhaps even more on the perception of distant weather and climates. It finds expression in the blue skies of travel posters (whether of the British Isles, the Oregon coast, or New Zealand) and in promotional literature meant to caress and nurture the utopian notions of cruise ship passengers (Porter and Lukermann 1976, pp. 214–15).

Both perceived and real knowledge of climate are reflected in the annual business cycles of tourist-oriented retailers in San Francisco, where the merchants of Fisherman's Wharf are busiest in 'summer' – the foggy season from May through August and into the generally warm and clear month of September or even October. By late spring, the milling crowds of tourists brave the frequently chilly and foggy weather and wander into the souvenir shops, art galleries, and eating places to buy mementos and experiences to be relived in conversations back home. That many visitors wear attire more in tune with Hawaii or Acapulco than with the wisps of fog swept in on the seabreeze, testifies to the unfortunate gap between the application of perception from afar – and the reluctance to come to terms with reality – and the knowledge of the actual climate. Lingering perceptions may be out of step with new conditions and 'imagination, distortion, and ignorance still embroider our private landscapes' (Lowenthal 1961, p. 249). Ever since the seventeenth century London has had a reputation of smoke and 'pea soup' fog. But, since the passage of the British Clean Air Act (1956) the situation has vastly improved (Neiburger et al. 1982, p. 351); now, London air quality is better than in some American cities at certain times of year. Climatology as it applies to the brightening or correction of these images can play an important role.

Annual meetings of the A.A.G., the American Meteorological Society, the American Geophysical Union, and Applied Geography Conferences are some of the marvelous sources of information and ideas for the applied atmospheric scientist. On account of their multidisciplinary nature, the meetings of the American Association for the Advancement of Science are, from my experience, particularly worthwhile, not only for the activities of the atmospheric and hydrospheric sciences (Section W), but of some other sections also (Lier 1983). The most recent national meeting (Chicago, February, 1987) included symposia entitled 'Climatology and Weather Information in Agriculture and Forestry,' 'Climatic Variability and Impacts on U.S. Water Resources,' 'Physical Problems in the Great Lakes,'[28] and 'Review of the 1977–87 Decade of Action to Combat Desertification.' It was disappointing that in these four symposia, all very pertinent to applied climatology, I was able to identify by title meteorologists, hydrologists, biologists, geologists, oceanographers, and engineers, but no geographers (Herschman 1987).

As I have argued elsewhere (Lier 1983), it seems that we miss too many chances to listen, compare notes, show our wares, and add our perspective.

Conclusion

The long history of applied geography, its continued independent existence in the nonacademic world, its pervasiveness in academic geography, and the reciprocity between theory and application have been important themes of this paper. The Applied Climatology section in particular shows that many studies have an applied aspect, but that climatology is by no means a unique part of geography. It follows that turning against the principle of applied geography would be like turning against geography as a whole. And, since I love geography for its excitement, intellectual scope, and satisfaction more than any other field, the principle is safe with me. But, as this study has demonstrated, in practice there are conceptually different kinds of applied geography, and here I must confess that I do not agree with them all. Not all kinds contribute equally to the base of scholarship, to conceptual and philosophical depth, to the crucial theoretical infrastructure of the profession. Most applied geography contributes *ceteris paribus* to the infrastructure, in direct proportion to the reciprocal relations discussed above. In many cases they develop spontaneously as with professor-student team work; elsewhere, encouragement seems to be needed, as with internships in off-campus employment of the kind represented by the *Directory of Applied Geographers*.

The protection and expansion of the infrastructure of academic geography is the main responsibility of academic geographers. This is true also of nonacademic geographers who practice geography, regardless of job title. Geographers should not get lost to their departments or to academia in general, but are under obligation to communicate their professional experiences regularly to the wider academic geographic community. The benefits will be mutual. Without everyone's concern for the infrastructure – if, for instance, we neglect it under the pressures of the moment – we shall soon 'have nothing to teach, nothing to apply, nothing to justify our continued existence' (Jordan 1987, p. 1). Obviously, at present it is more important to help our students to find employment outside academia than when academic jobs are plentiful; but, our measures must not be at the expense of our infrastructure and must take at least three principles into account:

(1)   we must remain masters in our own house in the sense that no form of support from the outside be made conditional on influence of departmental decision making (Lier 1984);

(2)   no faculty member must ever feel pressured to interrupt his or her research, reading, or other professional activities to participate in problem solving for the outside world; and

(3)   we must be very careful with the allocation of our resources.

Regarding the first two points, my relatively small department (seven full-time faculty) has done very well, though (should I admit it?), it has never been faced with the predicament of the first point. As to the allocation of resources, since student enrollments began to decline, as they did in many other (geography) departments, we have devoted more time and effort to improving our students' chances to find *Directory of Applied Geographers*-type employment, which has resulted in increased enrollments. In1975–76 the more 'applied' B.S. degree was added to the B.A. degree, and over the years about three times more students have opted for the B.S. Since 1978–79, the department has administered an environmental studies program (and has had its present name), although a course in environmental impact report writing is somewhat older; our majors have been about equally divided between geography and environmental studies. We have perhaps the best cartography facilities in central California and hope to improve them further in the near future; since 1985–86, we have offered a certificate in cartographic communication. Since 1969–70, we have had a graduate program leading to the M.A. degree, producing a number of applied theses on such topics as water balance forecasting, tourism, planning of health care facilities, and regional planning and development. Although its practical value to our department is undeniable, it is my view that the time is near for our faculty to start paying relatively less attention to nonacademic job training and more to the infrastructure. Once that is neglected – and in this case I do not suggest that it has been – it may, considering the rapid growth of theory and the limitation of resources, be impossible to rehabilitate it. In the case of a young department in a young university (such as ours) such neglect would be particularly serious. But, even departments, young or old, in universities of high repute have to be very cautious. Quite apart from any personal insistence on scholarship that arises from a faculty member's reasons for seeking an academic appointment initially, the unfortunate experience at Harvard teaches a very important pragmatic lesson: an excellent reputation of scholarship serves, if not as a guarantee, as a bulwark against internal or external personality conflict, low student enrollments, interdepartmental politics or fiscal problems (Smith 1987).

Notes

1. An excellent recent example is Parsons (1986).
2. Admittedly, not an easy task, but see, e.g., Simmons (1981) and Parsons (1986).
3. The first alliance, the California Geographic Alliance, was founded in 1984; the second, the Northern California Geographic Alliance, in 1985. There are now fifteen alliances, all strongly supported by the National Geographic Society. See also Gardner (1986) and Salter (1986).
4. For a recent discussion of dialectical relations, see, e.g., Colledge (1979).
5. For a critical examination of the theory-praxis relationship, see Habermas (1973); chapter 7 is devoted to contemporary conditions.
6. See also Hooson (1984) and several essays in Demko and Fuchs (1984).
7. Adler's paper, published in 1910, was abridged and translated from the Russian by H. de Hutorowics, whose name is cited under the title in Adler (1911). I am indebted to Sandra Marburg for suggesting these references.
8. For the recent invasion and other examples, see Neumann and Flohn (1987).
9. The *Geographical Review* began an applied geography section in October, 1976, but it

ended with the April, 1980 issue.
10. Excluding institutional and corporate members.
11. I am indebted to Colin Lear for assistance.
12. For general background on the intermittent calls for social relevance, see Johnston (1985) and Taylor (1985), and for views on the seventeenth century, Kuhn (1977, pp. 59, 115).
13. See also Buttimer (1974, p. 3).
14. See also William-Olsson (1983, pp. 162–63).
15. For experience, see Marcus (1982, pp. 84–89).
16. The Program is scheduled to absorb in late 1987 the World Ocean Circulation Experiment (WOCE), which stresses ocean-climate relations.
17. UNESCO publishes a series of MAB Technical Notes on a broad spectrum of applied geography with a strong Third World emphasis.
18. FitzRoy, a respected meteorologist, was appointed in 1854 to the newly created post ashore of Meteorological Statist; a few years later he founded the present Meteorological Office. For a recent synopsis of original material on Darwin's voyage of the *Beagle*, see Anonymous (1977). This note from pp. 25–26.
19. See also Anonymous (1981a), Barry (1981, pp. 290–92), Plasschaert (1969), and Price (1981, pp. 153–65, 274, 287).
20. See, e.g., Browning (1982).
21. Since the nuclear accident at Chernobyl on 26 April 1986, many articles and news items have appeared in *Science*. For a recent prognosis, see the 26 June 1987 issue (Wilson 1987). See also Strauss and Gros (1986), which shows the generally westward progression of radioactivity to May 1st (noon).
22. In Russian, though a translation has been published (Donehoo 1964). See also Löf *et al.* (1966) and Miller (1981).
23. See, e.g., Nix (1985). Important data bases are agroclimatic atlases, such as Thran and Broekhuizen (1965), Gol'tsberg (1972), and Baier (1976).
24. See also the climate impact studies by Le Houérou (1985) and Jodha and Mascarenhas (1985). Further examples can be gleaned from Thomas (1956).
25. See, e.g., Young (1987) and the reproductions in Muller and Oberlander (1978).
26. Carl Sauer viewed physical geographic knowledge as fundamental to discerning observation: 'The land talks to you when you have an understanding of what it is that you are standing on and looking at' (Hewes 1983, p. 144).
27. See, e.g., Parsons (1986, pp. 374–75).
28. See Changnon (1987).

Bibliography

Adler, B.F. 1911. Maps of primitive peoples. *Bulletin of the American Geographical Society* 43:669–79.

Anonymous 1977. *A Narrative of the Voyage of H.M.S. Beagle*. London: The Folio Society.

Anonymous 1981a. *Avalanche Atlas*. Paris: UNESCO.

Anonymous 1981b. *Directory of Applied Geographers*. Washington: Association of American Geographers.

Anonymous 1981c. *Meteorological Aspects of the Utilization of Solar Radiation as an Energy Source* (Technical Note No. 172). Geneva: World Meteorological Organization.

Anonymous 1982. *A.A.G. Directory 1982*. Washington: Association of American Geographers.

Anonymous 1987. *A.A.G. Directory 1987*. Washington: Association of American Geographers.

Bach, W. *et al.*, eds. 1981. *Food – Climate Interactions.* Boston: Reidel.
Baier, W., ed. 1976. *Agroclimatic Atlas, Canada.* Ottawa: Agriculture Canada.
Barry, R.G. 1981. *Mountain Weather and Climate.* London: Methuen.
Barry, R.G., and Perry, A.H. 1973. *Synoptic Climatology: Methods and Applications.* London: Methuen.
Beaujeu-Garnier, J. 1983. Autobiographical essay. In *The Practice of Geography*, ed. A. Buttimer, pp. 141–152. New York: Longman.
Berkofsky, L. and Wurtele, M.G., eds. 1986. *Progress in Desert Research.* Totowa, NJ: Rowman and Littlefield.
Berkofsky, L. *et al.* 1981. *Settling the Desert.* New York: Gordon and Breach.
Biswas, M.R. and Biswas, A.K., eds. 1979. *Food, Climate, and Man.* New York: Wiley.
Braak, C. 1925? Het klimaat van Nederlandsch-Indië (The climate of the Netherlands Indies), *Verhandelingen No. 8*, Koninklijk Magnetisch en Meteorologisch Observatorium, Batavia 1(2).
Brewster, N.A. 1987a. National Science Foundation. Telephone conversation, 9 March 1987.
Brewster, N.A. 1987b. Understanding planet earth: NSF funds for a long-term comprehensive study of earth systems, *Discovery* 2(4).
Briggs, D.J. 1981. Editorial, *Applied Geography* 1:1–8.
Brookfield, H.C. 1983. Experience of an outside man. In *Recollections of a Revolution: Geography as Spatial Science*, eds. M. Billinge *et al.*, pp. 27–38. New York: St. Martin's Press.
Browning, K. A., ed. 1982. *Nowcasting.* New York: Academic Press.
Budyko, M.I., ed. 1963. *Atlas Teplovogo Balansa Zemnogo Shara (Atlas of the Heat Balance of the Earth).* Moscow: Academia Nauk S.S.S.R.
Burton, I. et al. 1978. *The Environment as Hazard.* New York: Oxford University Press.
Buttimer, A. 1974. *Values in Geography.* Washington: Association of American Geographers. Commission on College Geography. Resource Paper No. 24.
Buttimer, A. 1979. Erewhon or nowhere land. In *Philosophy in Geography*, eds. S. Gale and G. Olsson, pp. 9–37. Boston: Reidel.
Buttimer, A. 1984. Ideal und Wirklichkeit in der angewandten Geographie, *Münchener Geographische Hefte* 51.
Byrne, R. et al. 1982. Recent rainfall trends on the margins of the subtropical deserts: a comparison of selected northern hemisphere regions, *Quaternary Research* 17:14–25.
Changnon, S.A. 1987. Climate fluctuations and record-high levels of lake Michigan, *Bulletin of the American Meteorological Society* 68:1394–1402.
Claval, P. 1984. Conclusion. In *Geography Since the Second World War: An International Survey*, eds. R.J. Johnston and P. Claval, pp. 282–289. Totowa, NJ: Barnes and Noble.
Colby, C.E. 1924. California raisin industry: a study in geographic interpretation, *Annals, Association of American Geographers* 14:49–100.
Davies, E. 1941. The pattern of transhumance in Europe, *Geography* 26:155–168.
Demko, G.J. and Fuchs, R.J., eds. 1984. *Geographical Studies on the Soviet Union: Essays in Honor of Chauncy D. Harris.* Chicago: University of Chicago, Department of Geography (Research Paper No. 211).
Denisov, G.V. and Elovskaia, L.G. 1979. *Oats in the Permafrost Zone.* Novosibirsk: Nauka, Sibirskoe otd – nie.
Dickenson, J.P. *et al.* 1983. *A Geography of the Third World.* New York: Methuen.
Dobel, R. 1968. *Lexikon der Goethe Zitate.* Zürich: Artemis Verlag.
Donehoo, I.A. 1964. *Guide to the Atlas of the Heat Balance of the Earth.* Washington: U.S. Department of Commerce, Weather Bureau.
Evans, E.E. 1940. Transhumance in Europe, *Geography* 25:172–180.
Eysberg, C.D.1984. *De Verenigde Staten.* Weesp (Neth.): Romen.
FAO, UNESCO, WMO. 1979? *World Map of Desertification* (A/CONF. 74/2).
Gade, D.W. 1978. Windbreaks in the lower Rhône valley, *Geographical Review* 68:127–144.

Gallais, J. 1984. *Hommes du Sahel: Espaces-Temps et Pouvoirs. Le Delta intérieur du Niger 1960–1980*. Paris: Flammarion.

Gardner, D.P. 1986. Geography in the school curriculum, *Annals, Association of American Geographers* 76:1–4.

Garnett, A. 1935. Insolation, topography, and settlement in the Alps, *Geographical Review* 25: 601–617.

Geertz, C. 1963. *Agricultural Involution: the Process of Ecological Change in Indonesia*. Berkeley: University of California Press.

Golledge, R.G. 1979. Reality, process, and the dialectical relation between man and environment. In *Philosophy in Geography*, eds. S. Gale and G. Olsson, pp. 109–120. Boston: Reidel.

Gol'tsberg, I.A., ed. 1972. *Agroklimaticheskiĭ Atlas Mira*. Moscow: Glavnoe Upravlenie Geodeziĭ i Kartografiĭ Pri Sovete Ministrov S.S.S.R., Gidrometeoizdat.

Goudie, A. 1986. *The Human Impact on the Natural Environment*. Cambridge, MA: MIT Press.

Granger, O.E.1980a. Climatic variations and the California raisin industry, *Geographical Review* 70:300–313.

Granger, O.E. 1980b. The impact of climatic variation on the yield of selected crops in three California counties, *Agricultural Meteorology* 22:367–386.

Gregory, D. 1978. *Ideology, Science and Human Geography*. London: Hutchinson.

Habermas, J. 1973. *Theory and Practice*. Boston: Beacon Press.

Hägerstrand, T. 1974. Commentaries on 'values in geography'. In *Values in Geography* by A. Buttimer, pp. 50–54. Washington: Association of American Geographers. Commission on College Geography. Resource Paper No. 24.

Hägerstrand, T. 1983. In search for the sources of concepts. In *The Practice of Geography*, ed. A. Buttimer, pp. 238–256. New York: Longman.

Haggett, P. 1983. *Geography: A Modern Synthesis*. New York: Harper and Row.

Hare, F.K. 1973. Energy-based climatology and its frontier with ecology. In *Directions in Geography*, ed. R.J. Chorley, pp. 171–192. London: Methuen.

Hare, F.K. 1977. Man's world and geographers: a secular sermon. In *Geographic Humanism, Analysis and Social Action: Proceedings of Symposia Celebrating a Half Century of Geography at Michigan*, eds. R. Deskins *et al.*, pp. 261–273. Ann Arbor: Department of Geography, University of Michigan.

Hare, F.K. 1979. Food, climate, and man. In *Food, Climate, and Man*, eds. M.R. Biswas and A.K. Biswas, pp. 1–11. New York: Wiley.

Hare, F.K. 1985. Climatic variability and change. In *Climate Impact Assessment: Studies of the Interaction of Climate and Society*, eds. R.W. Kates *et al.*, pp. 37–68. New York: Wiley.

Harris, C.D. 1985. In Memoriam, Stephen Bar Jones, 1903–1984, *Annals, Association of American Geographers* 75:271–276.

Harris, S.A. 1986. *The Permafrost Environment*. Totowa, NJ: Barnes and Noble.

Harte, J. 1983. An investigation of acid precipitation in Qinghai Province, China, *Atmospheric Environment* 17:403–408.

Herschman, A. 1987. *1987 AAAS Annual Meeting Program*. Washington: American Association for the Advancement of Science.

Hewes, L. 1983. Carl Sauer: a personal view, *Journal of Geography* 82:140–147.

Hobbs, J.E. 1980. *Applied Climatology*. Boulder, CO: Westview Press.

Holtgrieve, D.G. and Trevors, J. 1978. *The California Wine Atlas*. Hayward, CA: Ecumene Associates.

Hooson, D.J.M. 1984. The Soviet Union. In *Geography since the Second World War: an International Survey*, eds. R.J. Johnston and P. Claval, pp. 79–106. Totowa, NJ: Barnes and Noble.

Jäger, J. 1983. *Climate and Energy Systems: A Review of Their Interactions*. New York: Wiley.

James, P.E. 1967. On the origin and persistence of error in geography, *Annals, Association of American Geographers* 57:1–24.

Jodha, N.S. and Mascarenhas, A.C. 1985. Adjustment in self-provisioning societies. In *Climate Impact Assessment: Studies of the Interaction of Climate and Society*, eds. R.W. Kates et al., pp. 437–464. New York: Wiley.

Johnston, R.J. 1983a. *Geography and Geographers: Anglo-American Human Geography since 1945.* London: Edward Arnold.

Johnston, R.J. 1983b. *Philosophy and Human Geography.* London: Edward Arnold.

Johnston, R.J. 1985. Introduction: exploring the future of geography. In *The Future of Geography*, ed. R.J. Johnston, pp. 3–26. New York: Methuen.

Jordan, T.G. 1987. President's column – priorities, *AAG Newsletter* (Association of American Geographers) 22(7):1.

Karan, P.P. 1985. Geographers as consultants on UN projects, *Professional Geographer*, 37:470–473.

Kawasaki, T. 1985. Fisheries. In *Climate Impact Assessment: Studies of the Interaction of Climate and Society*, eds. R.W. Kates *et al.*, pp. 131–153. New York: Wiley.

Kuhn, T.S. 1970. *The Structure of Scientific Revolutions.* Chicago: University of Chicago Press.

Kuhn, T.S. 1977. *The Essential Tension: Selected Studies in Scientific Tradition and Change.* Chicago: University of Chicago Press.

Landsberg, H.E. 1981. *The Urban Climate.* New York: Academic Press.

Le Houérou, H.N. 1985. Pastoralism. In *Climate Impact Assessment: Studies of the Interaction of Climate and Society*, eds. R.W. Kates *et al.*, pp. 155–185. New York: Wiley.

Leighly, J. 1964. Charles Warren Thornthwaite, March 7, 1899–June 11, 1963, *Annals, Association of American Geographers* 54:615–621.

Leighly, J. 1983. Memory as mirror. In *The Practice of Geography*, ed. A. Buttimer, pp. 80–89. London: Longman.

Leng, G., ed. 1982. *Desertification: A Bibliography with Regional Emphasis on Africa.* Bremen: Schwerpunkt Geographie, Fachbereich 1, Universität Bremen.

Lettau, H. 1969. Evapotranspiration climatonomy. 1. A new approach to numerical prediction of monthly evapotranspiration, runoff, and soil moisture storage, *Monthly Weather Review* 97:691–699.

Lettau, H. and K. 1969. Shortwave radiation climatonomy, *Tellus* 21:208–222.

Lier, J. 1968. *The Impact of the Rural Economy on Urban Structure and Form in the Canadian Wheat Belt.* Unpublished Ph. D. thesis, Department of Geography, University of California, Berkeley.

Lier, J. 1971. Farm mechanization in Saskatchewan, *Tijdschrift voor Economische en Sociale Geografie* 62:180–89.

Lier, J. 1983. The reticent geographer, the popular image, and pre-university education: perceptions from recent interdisciplinary experience, *California Geographer* 23:1–13.

Lier, J. 1984. Comments on industry-government-academic cooperation, *Professional Geographer* 36:219–220.

Löf, G.O.G. et al. 1966. *World Distribution of Solar Radiation.* Madison: University of Wisconsin, Solar Energy Laboratory.

Lowenthal, D. 1961. Geography, experience, and imagination: towards a geographical epistemology, *Annals, Association of American Geographers* 51:241–260.

Lydolph, P.E. 1979. *Geography of the U.S.S.R.* Elkhart Lake, Wisconsin: Misty Valley Publishing.

Marcus, M.G. 1979. Coming full circle: physical geography in the·twentieth century, *Annals, Association of American Geographers* 69:521–532.

Marcus, M.G. 1982. (Interview). In *Conversations with Geographers: Career Pathways and Research Styles*, interlocutor C.E. Browning, pp. 75–91. Chapel Hill: University of North Carolina, Department of Geography (Studies in Geography No. 16).

Mather, J.R. 1974. *Climatology: Fundamentals and Applications.* New York: McGraw-Hill.
Miller, D.H. 1977. *Water at the Surface of the Earth: An Introduction to Ecosystem Hydrodynamics.* New York: Academic Press.
Miller, D.H. 1981. *Energy at the Surface of the Earth: An Introduction to the Energetics of Ecosystems.* New York: Academic Press.
Muller, R.A. and Oberlander, T.M. 1978. *Physical Geography Today: A Portrait of a Planet.* New York: Random House.
Neiburger, M. *et al.* 1982. *Understanding Our Atmospheric Environment.* San Francisco: Freeman.
Neumann, J. and Flohn, H. 1987. Great historical events that were significantly affected by the weather: Part 8. Germany's war on the Soviet Union, 1941–45. I. Long-range weather forecasts for 1941–42 and climatological studies, *Bulletin of the American Meteorological Society* 68:620–630.
Nieuwolt, S. 1977. *Tropical Climatology.* New York: Wiley.
Nix, H.A. 1985. Agriculture. In *Climate Impact Assessment: Studies of the Interaction of Climate and Society,* eds. R.W. Kates et al., pp. 105–130. New York: Wiley.
Oke, T.R. 1976. Inadvertent modification of the city atmosphere and the prospects for planned urban climates, *Proceedings of the Symposium on Meteorology Related to Urban and Regional Land-Use Planning, Asheville, N.C.* Geneva: World Meteorological Organization.
Oke, T.R. 1978. *Boundary Layer Climates.* London: Methuen.
Oliver, J.E. 1977. *Perspectives on Applied Physical Geography.* North Scituate, MA: Duxbury Press.
Oliver, J.E. 1979. *Physical Geography: Principles and Applications.* North Scituate, MA: Duxbury Press.
Omernik, J.M. 1987. Map supplement – ecoregions of the conterminous United States, *Annals, Association of American Geographers* 77:118–125.
Parsons, J.J. 1986. A geographer looks at the San Joaquin valley, *Geographical Review* 76: 371–389.
Parsons, J.J., and Vonnegut, N., eds. 1983. *60 Years of Berkeley Geography 1923–1983.* Berkeley: University of California, Department of Geography.
Peters, G.L. 1984. Trends in California viticulture, *Geographical Review* 74:455–467.
Peters, G.L. 1987. The emergence of regional cultivar specializations in California viticulture, *Professional Geographer* 39:287–297.
Phlippomeau, M. 1981. The rise of applied geography, *International Social Science Journal* 33:131–159.
Pilkington, R. 1970. *Small Boat on the Lower Rhine.* London: Macmillan.
Plasschaert, J.H.M. 1969. Weather and avalanches, *Weather* 24:99–102.
Porter, P.W. 1977. Commentary. In *Geographic Humanism, Analysis and Social Action: Proceedings of Symposia Celebrating a Half Century of Geography at Michigan,* eds. R. Deskins *et al.,* pp. 275–289. Ann Arbor: Department of Geography, University of Michigan.
Porter, P.W. and Lukermann, F.E. 1976. The geography of utopia. In *Geographies of the Mind,* eds. D. Lowenthal and M.J. Bowden, pp. 197–223. New York: Oxford University Press.
Pred, A. 1983. From here and now to there and then: some notes on diffusions, defusions and disillusions. In *Recollections of a Revolution: Geography as Spatial Science,* eds. M. Billinge *et al.,* pp. 86–103. New York: St. Martin's Press.
Price, L.W. 1981. *Mountains and Man.* Berkeley: University of California Press.
Reitsma, H.A. and Kleinpenning, J.M.G. 1985. *The Third World in Perspective.* Assen (Neth.): Van Gorcum.
Richter, M. and Schmiedecken, W. 1985. The climate of cases and its ecological significance, *Erdkunde* 39:179–197.

Riebsame, W.E. 1985. Research in climate-society interaction. In *Climate Impact Assessment: Studies of the Interaction of Climate and Society*, eds. R.W. Kates *et al.*, pp. 69–84. New York: Wiley.

Sack, R.D. 1980. *Conceptions of Space in Social Thought: a Geographic Perspective.* Minneapolis: University of Minnesota Press.

Salomé, A.I. 1979. Sneeuwlawines in de Alpen, *Geografisch Tijdschrift*, Nieuwe Reeks 13:286–298.

Salter, C.L. 1986. Geography and California's educational reform: one approach to a common cause, *Annals, Association of American Geographers* 76:5–17.

Sawatzky, H.L. 1987. Legacy and stewardship. In *Carl O. Sauer – A Tribute*, ed. M.S. Kenzer, pp. 205–13. Corvallis: Oregon State University Press.

Shlemon, R.J. 1979. Applications of physical geography: quaternary, soil-stratigraphy, and geomorphology, *Geographical Survey* 8:9–17.

Simmons, I.G. 1981. *The Ecology of Natural Resources.* New York: Wiley.

Smith, N. 1987. Academic war over the field of geography: the elimination of geography at Harvard, 1947–1951, *Annals, Association of American Geographers* 77:155–172.

Spink, J. and Moodie, D.W. 1972. *Eskimo Maps from the Canadian Eastern Arctic.* Toronto: B.V. Gutsell, Department of Geography, York University (Monograph No. 5).

Staple, W.J. et al. 1960. Conservation of soil moisture from fall and winter precipitation, *Canadian Journal of Soil Science* 40:80–88.

Stoddart, D.R. 1986. *On Geography and Its History.* Oxford: Basil Blackwell.

Strauss, B. and Gros, J.M. 1986. Information sur l'Accident de Tchernobyl, *La Météorologie*, série 7, no. 15:14–15.

Taylor, P.J. 1985. The value of a geographical perspective. In *The Future of Geography*, ed. R.J. Johnston, pp. 92–110. New York: Methuen.

Thomas, W.L. Jr., ed. 1956. *Man's Role in Changing the Face of the Earth.* Chicago: University of Chicago Press.

Thompson, R.D. 1974. Climate and permafrost in Canada, *Weather* 29: 298–305.

Thran, P. and Broekhuizen, S. eds. 1965. *Agro-Climatic Atlas of Europe.* Amsterdam: Elsevier Publishing Company.

Vale, T.R., ed. 1986. *Program Against Growth: Daniel B. Luten on the American Landscape.* New York: Guilford Press.

Van der Stok, J.P. 1897. *Wind and Weather, Currents, Tides and Tidal Streams in the East Indian Archipelago.* Batavia: Government Printing Office.

Walter, H. and Breckle, S.-W. 1985. *Ecological Systems of the Geobiosphere*, 2 Vols. New York: Springer-Verlag.

Walter, H., and Lieth, H. 1960–1967. *Klimadiagramm-Weltatlas.* Jena: VEB Gustav-Fisher-Verlag.

Watts, M. 1983. *Silent Violence: Food, Famine and Peasantry in Northern Nigeria.* Berkeley: University of California Press.

White, G.F. 1973. Natural hazards research. In *Directions in Geography*, ed. R.J. Chorley, pp. 193–216. London: Methuen.

Whyte, A.V. and Burton, I., eds. 1980. *Environmental Risk Assessment.* New York: Wiley.

Wilken, G.C. 1972. Microclimate management by traditional farmers, *Geographical Review* 62:544–560.

William-Olsson, W. 1983. My responsibility and my joy. In *The Practice of Geography*, ed. A. Buttimer, pp. 153–66. New York: Longmans.

Williams, F.L. 1963. *Matthew Fontaine Maury: Scientist of the Sea.* New Brunswick, NJ: Rutgers University Press.

Willis, W.O. and Carlson, C.W. 1962. Conservation of winter precipitation in the northern plains, *Journal of Soil and Water Conservation* 17:122–123, 128.

Wilson, R. 1987. A visit to Chernobyl, *Science* 236:1636–40.

Wooster, W.S. and Fluharty, D.L., eds. 1985. *El Niño North: Niño Effects on the Eastern Subarctic Pacific Ocean.* Seattle: Washington Sea Grant Program, University of

Washington.

Wright, J.K. 1966. *Human Nature in Geography*. Cambridge, MA: Harvard University Press.

Young, J.R.C. 1987. Cover: Painting by Jacob van Ruisdael (1628–92), *Weather* 42:308.

John Lier
Department of Geography and
Environmental Studies
California State University
Hayward, CA 94542
U.S.A.

7. Current Status, Trends, and Problem-Solving in Applied Climatology

Applied Studies

A liberal education is one that presumably provides a variety of experiences for dealing with the vagaries of life. Many hold that the more varied these experiences, the greater will be a person's ability to deal with technological and social changes and his or her potential for contributing to society. In contrast, some see an evolving trend, fostered by the perceived, increasingly complex nature of society (as well as student enrollment trends), which must necessarily lead to fewer experiences, more prescribed learning within a chosen subject, and, again presumably, greater competence in specific aspects of a discipline. Under this viewpoint, application of specific knowledge could have a potentially greater impact on society than a general education might provide. The purpose of this essay is to comment briefly on these approaches to education in geography, turn the focus to applied climatology, and, finally, provide some typical examples of problems applied climatologists often work on as employees of private and public agencies.

Under the first viewpoint, geography would be a traditional liberal arts subject among many studied at the undergraduate and graduate levels; some limited specialization within the discipline would be possible, but not at the expense of exposure to portions of the entire field. Future application of geographical principles would be up to the student. The second route would allow very little deviation from a set curriculum established around topical areas. Under this scenario, geographers would be instructed in the methodologies, principles, and philosophies of their discipline with an eye toward practical applications in a variety of real world settings. Such emphasis would also produce graduates who would find employment as topical specialists – e.g., as cartographers, rather than as geographers.

The debate between traditional and applied educational experiences began in the 1960s when educational relevance became a national issue. Although geography is used here as the example, similar trends were mirrored in other fields, ecology for example. The 1969 National Environmental Policy Act required preparation of Environmental Impact Statements prior to modification, real or imagined, of ecosystems. The Act forced application of ecological principles to describe various environments and to evaluate which actions need be considered to mitigate potential damage. Various texts and manuals appeared to support such description and analysis, a number written by people working for private firms (e.g., De Santo 1978; Gilbertson et al. (1985). Even engineering, probably the best example of applied science, was not immune.

M. S. Kenzer (ed.), Applied Geography: Issues, Questions, and Concerns, 99–113.
© 1989 *Kluwer Academic Publishers.*

Geographers obviously were carried along with national trends and, by the late 1970s, were very much involved in intradisciplinary review. For example, Michigan State University organized a national conference in which applied – 'versus' would serve as a more correct word for some attendees – and academic geography were debated among people from the private sector, governmental agencies, and academe (Winters and Winters 1977). The conference marked a time when the role of applied studies within geography overall was still open to question. Harrison and Larsen (1977) offered this comment:

The real world is operationally based. The emphasis is on the individual to demonstrate his effectiveness on the job. The many geographers who have made significant professional contributions have done so without worrying over geography's proper role. Morrill, then, may be correct when he points out that in many cases a geographer's training may be inadequate to gain operational competency (p. 143).

The national debate among geographers appeared to reach a high water mark in the early 1980s, based on an examination of the number of kindred articles in the *Professional Geographer*. An applied geography interest group had already been established within the Association of American Geographers (AAG), by then a journal had appeared with the word applied in its title, and several topical conferences had been held.

Discussion about the role of applied studies and its place in geography is obviously not over – e.g., the volume you are reading. For example, Kenzer (1984), an avowed newcomer to the debate, states that applied studies appear to be a subset of human geography. Perhaps not (at least not where climatology is concerned), as we shall see.

Applied Climatology

Some Relationships and Definitions

Climatology is one subfield of physical geography. Strong ties are maintained between physical geography and environmental studies, remote sensing, quantitative methods, resource management, biogeography, and its most direct counterpart, meteorology. Climatology deals with the means, extremes, and explanations of patterns among the five weather elements: temperature, moisture, radiation, pressure, wind speed, and wind direction. The pattern of weather elements is, in turn, intimately linked to variations in the earth's surface; the link produces marked changes in heat, moisture, and momentum exchanges among the atmosphere, lithosphere, and hydrosphere.

The association with weather actually would place climatology as a subfield under meteorology, whereas its emphasis on location on the earth's surface would place it within physical geography. Given that both empirical observation of specific weather elements at a location or locations is necessary, combined with meteorological theory for explanation, climatology does not lie wholly within meteorology or geography. Springer (1972) suggests, for example, that climatology is an applied science, whose methods are strictly meteorological, but whose aims and results are geographical.

Subfields and Approaches

Climatology is often divided into four major subgroups (Fig. 1; Oliver 1981). Climatography involves the presentation of climatic data through various media. Examples range from a typical map of world climatic patterns based on selected criteria, to depiction of local-scale, three-dimensional arrays of temperature and moisture patterns in a valley. Physical climatology treats global and local energy and water balance regimes of the earth and atmosphere. Dynamic climatology is directed toward the study and explanation of atmospheric circulation over a large

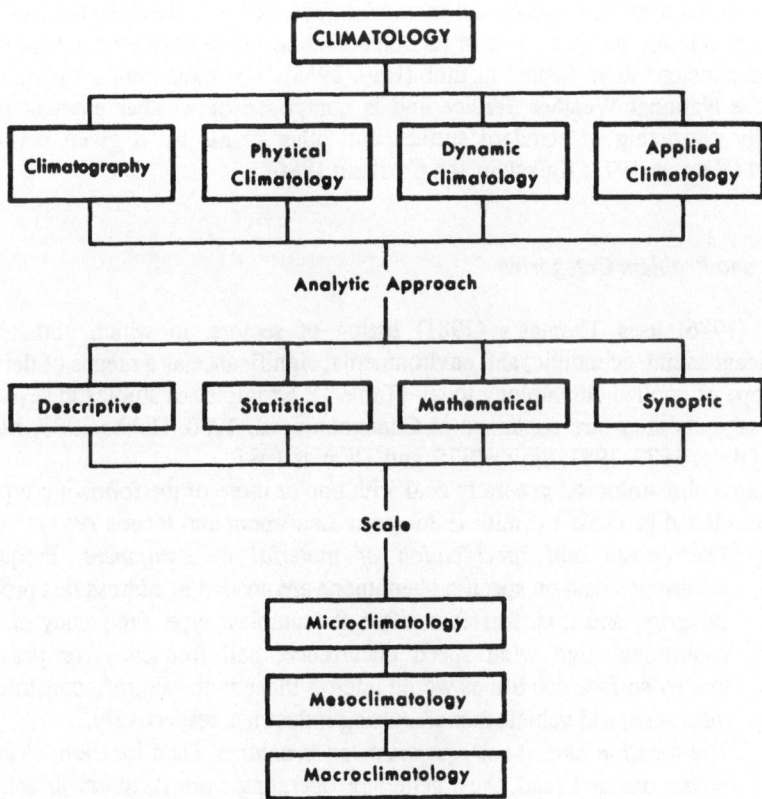

Fig. 1. Subgroups, analytic methods, and scale of climatic studies.

part of the earth's surface in terms of the available sources and transformation of energy. Applied climatology is the scientific use of climatic data and theoretical constructs for the solution of particular problems.

Oliver (1981) also shows the analytical approaches and scales that climatologists use in applied research. Descriptive climatologies are those most frequently encountered by the layperson; they are intended to be as concise, clear, illustrative, and as useful as possible. One of the more interesting examples of this genre is a recent explanation of US climatic conditions contained in *Places Rated Almanac* (Boyer and Savageau 1981). The text and graphs are intended to allow readers to compare weather elements across space and time as part of a decision matrix on the quality of life in major metropolitan areas. One could also argue that the result of Boyer and Savageau's work is a good application of climatological principles.

Statistical and mathematical approaches to climatological data are becoming the norm as our knowledge of climatic conditions and variations grows. The modeling of particular features of a given climate or climatic ensemble is often the result of these efforts (Cerveny and Balling 1984; Shelton 1985). The synoptic approach is not as common as other analytical methodologies, yet it is the only approach for certain problems. Its value lies in its consideration of weather element arrays in three dimensions at an instant in time (Hare 1955). The basic data set is obtained from the National Weather Service and is comprised of weather element maps, typically consisting of standard surface and other charts for a given period of interest (Harman 1971; Kalkstein and Corrigan 1986).

Scope and Problem Categories

Smith (1986) uses Thomas's (1981) listing of sectors, in which climate has significant social, economic, and environmental significance as a means of defining the scope of applied climatology today (Table 1). Examples of studies in several of these subjects areas are contained in Changnon et al. 1980, Hobbs 1980, Mather 1974, Oliver 1973, 1981, Smith 1975, and Thomas 1981.

Applied climatologists generally deal with one or more of the following types of problems listed in Table 1 (Smith 1986, citing Landsberg and Jacobs 1951):

1.   *The design and specification of material or equipment.* Frequency occurrence data on specific phenomena are needed to address this problem category, and it is, therefore, often the simplest type. Frequency of icing conditions, high wind speed occurrence, hail frequency, or prevalent muddy surface conditions would interest those in the aircraft, construction, insurance, and vehicle manufacturing industries, respectively.

2.   *The location and use of equipment or structures.* Dam location, electrical power demand (and, thus, generator operating cycles), reservoir scheduling, energy availability for passive solar heating, potential wind power installation sites, and advantageous, clear sky conditions for optical telescope location serve as examples.

Table 1. Sectors and activities where climate has significant social, economic, and environmental significance.

Primary Sectors	General Activities	Specific Activities
Food	Agriculture	Land use, crop scheduling and operations, hazard control, productivity, livestock and irrigation, pests and disease, soil tractionability
	Fisheries	Management, operations, yield
Water	Water disasters	Floods, droughts, pollution abatement
	Water resources	Engineering design, supply, operations
Health and Community	Human biometeorology	Health, disease, morbidity and mortality
	Human comfort	Settlement design, heating and ventilation, clothing, acclimatization
	Air pollution	Potential, dispersion, control
	Tourism and recreation	Sites, facilities, equipment, marketing, sports activities
Energy	Fossil fuels	Distribution, utilization, conservation
	Renewable resources	Solar, wind, water-power development
Industry and Trade	Building and construction	Sites, design, performance, operations, safety
	Communications	Engineering design, construction
	Forestry	Regeneration, productivity, biological hazards, fire
	Transportation	Air, water and land-facilities, scheduling, operations, safety
	Commerce	Plant operations, product design, storage of materials, sales planning, absenteeism, accidents
	Services	Finance, law, insurance, sales

Source: After Thomas (1981).

3. *The planning of a particular operation.* A higher order of complexity is represented here because a series of climatically sensitive processes must be defined and characterized. The scheduling of transportation related activities, such as fuel loads for aircraft at different times of the year, or delivery of materials to Alaskan north slope oil operations, would constitute typical problems.

4. *Climatic influences on biological activities.* The most complex of all such problems require an understanding of both physical and biological systems and the relationship of climatic measures to activities in each category. Examples encompass climatic fluctuations that have immediate conse-

quences on food supplies and living conditions (e.g., yield predictions, severe droughts, and extreme winters), and the climatic consequences of energy usage (e.g., carbon dioxide effects on planetary temperatures, nuclear winter, and stratospheric ozone depletion from possible chlorofluorocarbons effects).

The Past

Applied climatology enjoys a rich history in the modern era, extending back to the early part of this century (Smith 1986). The development of new instrumentation in the 1930s allowed rapid construction of surface and some upper-air data bases for the continental US and for portions of Europe. Military needs during World War II added urgency to data collection and analysis and to the understanding of global climatic regimes (Miller 1987). Discovery in the 1940s and early 1950s of the Polar Front jet stream's role in atmospheric and storm circulation led to increased efforts to construct dynamic climatologies for various areas.

A gradual refinement of techniques in the 1950s and early 1960s, along with the emergence of statistical and mathematical approaches, led to new applications and extensions of climatological principles. Books, such as those by Chang (1968), Mather (1974), Oliver (1977), Olgyay (1963), Springer (1972), and Yoshino (1975), numerous contributions in a variety of journals, and the development of a national climate program plan in the late 1970s served to illustrate applications and to emphasize the importance of a climatological approach to various atmosphere-related problems. Indeed, one reason why climatologists have been involved only peripherally, at best, in the debate on the necessity to introduce applications into the geography curriculum is that applications are, by definition, part of the climatological approach and have been demonstrated repeatedly over the last four decades.

Current Situation

How is applied climatology faring as part of the educational activities called geography? In general, very well. Several items prove the point. About one-half of the designated State Climatologists are geographers. Historically, about one out of twelve articles in the *Journal of Applied Meteorology* – one of five periodicals published by the American Meteorological Society (AMS) – has been by individuals specifically claiming ties with geography. The percentage of geography-affiliated authors rose to seventeen in 1986, a reflection of growing interest in climatological subject matter. The AMS journal mentioned above was renamed in 1983 to include 'climate' in the title to acknowledge that interest, and an entirely new journal (*Journal of Climate*) is being sponsored by the society. Many more outlets are also available for publishing climatological research (Table 2).

About 12 percent of the positions advertised in *Jobs in Geography* are for climatologists of one or another interest area; the figure rises to 35 percent for positions advertised in *Employment Opportunities* published by the American Meteorological Society. About sixty of the 348 climatologists listed in the international *Directory of Climatologists* work outside of academia. The number of Certified Consulting Meteorologists (CCMs), a program run by the AMS, continues to increase, and many CCMs are climatologists. The number of firms that do consulting work in the atmospheric sciences has also increased, as a quick perusal of the *Bulletin* of the American Meteorological Society would show; more than one-half of the companies advertise for expertise in climatology, a percentage that has steadily risen over the last decade.

Table 2. A partial listing of journals containing climatological articles.

Monthly Weather Review
The Environmental Professional
Environmental Management
Journal of Climatology
Journal of Climate and Applied Meteorology
Annals, Association of American Geographers
Applied Geography
International Journal of Biometeorology
Agricultural Meteorology
Boundary Layer Meteorology
Environmental Science and Technology
Journal of the Air Pollution Control Association
Solar Age
Water Resources Research
Physical Geography
Mazingera
Ambio
Journal of Hydrology
Air, Water, and Soil Pollution
Environmental Conservation
Journal of Biogeography
Climatic Change
Atmospheric Environment
Journal of Soil and Water Conservation
Forest Science
Science
Nature
Journal of Environmental Quality

Curiously, the field and its application are relatively obscure as far as the whole of geography is concerned. Russell (1983) used the *Directory of Applied Geographers* to determine that only about 6 percent of the AAG membership claims ties with climatology. A discouraging note is that student membership in the climatology specialty group has dropped from about 30 percent to about 20 percent over the last three years, according to figures published in respective issues of the *AAG Newsletter*. There were no climatology papers presented at either of the first two Applied Geography Conferences, but several have been given recently. So far as I can determine, by examining individual issues, no articles on applied climatology have appeared in the *Geographical Review* or *Economic Geography*. Perhaps surprisingly, I found only three climatological papers in *Applied Geography* in the last five years (Archibold and Crisp 1983; Druyan, Goldreich, and Maximov 1986; Tarrant 1984), while four papers with a distinctly applied theme have appeared in the *Annals* of the Association of American Geographers over the same time period (Granger 1983; Greenland and Yorty 1985; Kalkstein and Corrigan 1986; Shelton 1985).

One could speculate at length why applied climatological articles do not appear regularly in geographical journals, but two reasons appear paramount. The first has to do with historical subdisciplinary ties to meteorology among climatologists. The 1940s and 1950s marked a period of close cooperation between the two fields, and it was not uncommon to find meteorologists in geography departments; several examples come to mind, including the universities of Wisconsin and Kansas. The reverse was also true – e.g., my Ph.D. adviser, a geographer, was a staff member in meteorology at Wisconsin. As meteorology emerged as a discipline in its own right in the 1950s, cooperation between the two subjects was retained within universities and at the national level through the AMS and governmental agencies. Furthermore, climatology students often take substantial coursework, when available, in meteorology; the membership on master's and doctoral committees, the research topics tackled by students, faculty, and professionals in public and private sectors, and the availability of a large applied climatology interest group in the AMS – most climatologists are AMS members, but very few applied climatologists are AAG members – reflect continued kinship. In light of this background, it is natural that climatologists gravitate toward outlets and associations outside of geography.

The second reason appears partly to reflect the nature of many applied climatological efforts themselves. As Smith (1986) points out, applied climatology often can be separated from other approaches by its emphasis on the needs of the customer or client, rather than the aims of the academic. This means that the applied research process often involves proprietary material, deals with an immediate problem (not a class of subject matter), represents application of known techniques, and typically does not utilize extentions of advanced or theoretical constructs. Papers or reports produced as a result of such a process often disappear into corporate or public entity coffers, because the motivation is not necessarily to publish, but to perform a service well so that further efforts can be funded.

Programs and Employment

There is no doubt that more nonthesis, topically oriented master's programs and baccalaureate degree tracks are being established in a number of fields including geography. Such programs, both by nature and design, emphasize applied aspects of the discipline. They are often specifically intended for those who may not wish to continue academic work at a higher level – at least not until after a period of employment. Many of these nonthesis and track programs are paradoxical.

Agencies and firms want practitioners who can do useful research in more than one area; often this means command of a broader spectrum of knowledge as well as special skills (Marotz 1983). Therefore, the applied geographer, particularly one who has had an internship or wishes to play a more complete role for an employer after graduation, takes courses in a variety of subjects (frequently from other disciplines), in addition to meeting a fixed set of topical requirements. Thus, an apparent limit on educational diversity, the antithesis of the traditional liberal arts education, often produces diversity because of the very real and practical demands placed on the applied scientist in today's society.

The rise in graduate interest in applied work generally is leading to two other clearly defined trends that will increase with time: the need to monitor past graduates to determine which changes are needed in a curriculum (both to assess how well they are doing in such positions, and to identify current and future jobs students may be interested in); and a widely expanded awareness of current openings in a variety of nontraditional establishments. People with skills and interests in applied climatology will continue to seek employment predominantly in state or federal agencies, or, increasingly, with private firms interested in their skills. However, unlike traditional academic staff positions, which are heavily advertised in traditional outlets, jobs in agencies and firms appear in entirely different places. The onus is on the potential practitioner to decipher listings and to prepare resume materials.

Examples

Four examples, briefly outlined below, end this chapter and serve as an overview of typical publicly and privately funded projects addressed under the rubric of applied climatology. They were selected to illustrate the scope, time and space scales, interrelationships among climatic and other factors, and range of studies under the categories as listed on Table 1. They also provide examples of objectives commonly employed in applied studies, including the development of procedures to solve generic problems where no direct application is illustrated, the application of a standard principle to solve a specific, generic problem, and the application of accepted principles to solve a specific problem for a specific client. The citations listed at the end of each example refer to additional studies that fall within the same category as the example.

Primary Sector: Water

An Irrigation Plan. – Golf courses are notorious consumers of water. Consumption rises exponentially in drier areas of the country, such as the deserts of the American southwest. Water costs in the Phoenix, Arizona area average $100,000 per year for the average, 100-acre course; large, exclusive courses average about double that figure, while very plush courses spend as much as $1,000,000 per year to keep playing surfaces immaculate. Most (75 percent) pump water from an on-site well; others draw from municipal supplies. Companies that supply irrigation equipment must be able to plan irrigation networks that cover all acreage within the confines of the course, but they have a competitive edge if they can also sell the client an irrigation scheduling scheme that minimizes water usage while keeping the grass in a healthy, playable condition.

A large vendor of sprinkler irrigation equipment was the contractor for a Phoenix golf course. The company subcontracted the development of a microcomputer-based irrigation scheduling model to a private consultant specializing in water balance problems. The consultant developed a scheme based on traditional water balance models. Input data included the area sprinkled by each sprinkler, the sprinkler output rate, the on-off cycle time, soil moisture infiltration – minimal because most soils have a caliche layer – soil moisture storage (assumed to be 2.5cm), and the actual and potential evaporation. The model output provided, among other items, the on-off sprinkler cycle time necessary for various irrigation rates. The irrigation supply firm purchased all rights to the software from the consultant and now sells it to other courses as a management tool (Mather 1978; Shelton 1983).

Primary Sector: Energy

Solar Collector Siting. – The design and testing of solar panels is usually done by engineers; the determination of energy available on site is often the province of climatologists. The task is not a trivial one, but it represents a straightforward application of earth sun geometrical relationships and estimates of atmospheric transmittance values.

Buildings and residential housing contractors often attempt to add features to dwellings (or have them available as an extra-cost option) that increase a structure's salability. Addition of solar space and water heating capabilities has become, and remains, a popular option in many parts of the country, despite the lapse of tax credits in 1985.

A contractor in the Kansas City area asked for a design package that would allow incorporation of passive solar heating to standard housing models planned for construction in a new subdivision. Availability of solar-driven features was to be heavily promoted as a sales incentive.

A consultant had to derive a series of tables and diagrams that would allow the contractor to site properly standard solar panels with known output characteristics. The task utilized a modification of procedures outlined by Ametek, Incorporated (1984) and Wilmott (1982). Wilmott, a geographer at the University of Delaware, has derived a model for estimating total global irradiance on a collector surface for any tilt and azimuth angle for any hour, day, or longer period. The model also allows the effect of site obstructions to be calculated so that panel setting can be adjusted optimally. Results from the model were combined with data from a worksheet that included additional climatic variables that could affect collector output, principally wind speed and temperature. The final product consisted of optimal siting characteristics for the intended panel usage by the contractor (Aguado 1986; Andersson *et al.* 1986; Diab and Garstang 1984).

Primary Sector: Health and Community

Air Quality. – Air quality in mountainous areas has traditionally been considered excellent. Increasing, and increasingly concentrated, recreational use has caused substantial deterioration of pristine atmospheric conditions in many sites in the montane western US. For example, Interstate-70 provides a convenient and fast route to ski areas in the Keystone-Breckenridge-Copper Mountain region of central Colorado. Vehicular traffic averages 500 units per hour during the tourist season (which now extends almost year around); backups of ten miles are often experienced at Eisenhower Tunnel, the main gateway to the recreational area from Denver, and populations of the three principal ski areas can climb into the tens of thousands on a given day. All of this activity leads to substantial air pollution, which causes marked decreases in visibility. The 1979 Clean Air Act is, in part, concerned with such visibility degradation, as are the inhabitants and users of mountainous regions, simply because the vistas that people expect can become less than expected.

Concern over pollution from woodstoves, the principal heat source during winter, led to a community request for assessment of its potential effect on visibility; ordinances were possible if a consultant's work showed that the noted substantial 'haziness' in the air probably resulted from wood-burning stove effluent.

The problem represents a typical application of climatological dispersion modeling with the added influence of topographic effects on dispersion patterns. The municipality needed a historical emission trend analysis, with separate pollution estimates calculated for autos and trucks, stationary sources other than woodstoves, and woodstoves themselves. An emissions inventory was created for each category. The results were combined with climatological data gathered from ski areas and from sources within the local community, as inputs into a modified climatological dispersion model (Greenland 1979). Results were compared with values from an air quality index developed for fireplace density-air quality standards assessment (Fosberg and Fox 1976). Although numerous methodologies are being developed continually by private and public agencies to assess the effects of woodburning, this approach yielded estimates that could be used to determine if a potential problem existed; further studies could then be initiated (De Freitas *et al.* 1985; Greenland and Yorty 1985; Kitada *et al.* 1986; Muller and Jackson 1985; Muschett 1981).

Primary Sector: Food

Corn Yields and Climatic Factors. – The association between climatic factors and agricultural output is a classic problem in climatology. Predicting yields has challenged agronomists, soil scientists, meteorologists, and climatologists for decades. The result of such prediction efforts are useful to agricultural economists, the international community, commodity traders, and lending agencies among others.

A typical approach is to use linear or curvalinear regression techniques to relate some weather element or elements observed over the growing season to a varietal yield, given certain soil types, soil moisture content, and other factors. The difficulty with the approach is that weather inputs consist only of long-term normals, and plants are characterized only by historical yields. As a result, the precision, accuracy, and error of this procedure suggest that alternatives might prove attractive.

The US Department of Agriculture and others have developed 'plant process' models – e.g., the Crop Estimation through Resource Environmental Synthesis model, or CERES-Maize – in which influential plant growth factors are characterized. Normal weather elements can also be synthesized into a simulated daily weather ensemble for the period of record.

A state agency funded a project to test the latter alternative approach. The research team selected a beginning year for the weather simulation, and four weather elements for each day of the year up to the corn sowing date, for placement into the weather data base. From the sowing to maturity date a different procedure was used, one which emphasized maximum inter- and intracorrelation among the four variables in order to replicate a realistic daily weather sequence. Plant process model inputs included such items as corn maturity type, photo-period sensitivity; duration and rate of grain filling, and grain number. Management information included sowing date, plant depth and plant density. The accuracy of the alternative method proved to be better than the regression approach under several simulated test conditions. Further work is being supported to extend the results to longer historical records and other regions of the country (Duchon 1986; Hubbard and Hanks 1983).

Summary

There is much discussion in academe today about the appropriate breadth and depth of the university experience, and how this experience should be gained. Some lean toward a liberal sprinkling of courses, while some favor a narrower range of choices and an emphasis on problem-solving through the use of real world examples. Geography is somewhat of a latecomer to the discussion, but it is currently attempting to deal with the liberal versus applied controversy.

Climatology as a subfield of geography utilizes various approaches to its subject matter, one of which has always emphasized application of known principles to solve practical problems. The problem-solving tradition, coupled with climatology's strong philosophical and educational ties with meteorology, has led to less controversy among climatologists about the applied-liberal debate. Applied climatology is doing rather well as a subject area overall, as evidenced by ever increasing outlets for research results, by the professional positions held by individuals claiming ties to the discipline, and by the number of employment opportunities available. Its status and growth within geography is less clear and likely to remain so, based on a sampling of AAG interest group affinities and the number of papers in traditional geographic periodicals. It is clear that applied climatologists continue to work on diverse problems in public and private sectors.

References

Aguado, E. 1986. Local-scale variability of daily solar radiation – San Diego County, California. *Journal of Climate and Applied Meteorology* 25:672–678.

Ametek, Inc. 1984. *Solar Energy Handbook.* Second Edition. Radnor, PA: Chilton Book Co.

Andersson, B., W. Carroll and M. Martin. 1986. Aggregation of US population centers using climate parameters related to building energy use. *Journal of Climate and Applied Meteorology* 25:596–614.

Archibold, O.W. and P.T. Crisp. 1983. The distribution of airborne metals in the Illawarra Region, New South Wales, Australia. *Applied Geography* 3:331–334.

Boyer, R. and D. Savageau. 1981. *Places Rated Almanac.* Chicago: Rand McNally Publishing Co.

Cerveny, R. and R. Balling. 1984. CONSTABLE: a simple one-dimensional model for climatologists in geography. *Professional Geographer* 36:188–195.

Chang, J. 1968. *Climate and Agriculture.* Chicago: Aldine Publishing Co.

Changnon, S., H. Critchfield, R. Durrenberger, C. Hosler and T. McKee. 1980. Examples of applications of climatic data and information provided by state climate groups. *Bulletin, American Meteorological Society* 61:1567–1569.

De Freitas, C., N. Dawson, A. Young, and W. Mackey. 1985. Microclimate and heat stress of runners in mass participation events. *Journal of Climate and Applied Meteorology* 24:184–191.

De Santo, R. 1978. *Concepts of Applied Ecology.* New York: Springer-Verlag, Inc.

Diab, R. and M. Garstang. 1984. Assessment of wind power potential for two contrasting coastlines of South Africa using a numerical model. *Journal of Climate and Applied Meteorology* 23:1645–1659.

Druyan, L.M., Y. Goldreich, and Z. Maximov. 1986. Wind energy in the Negev (Israel). *Applied Geography* 3:241–254.

Duchon, C. 1986. Corn yield prediction using climatology. *Journal of Climate and Applied Meteorology* 25:581–590.

Fosberg, M. and D. Fox. 1976. An air quality index for aid in determining mountain land use. *Proceedings of the Fourth National Conference on Fire and Forest Meteorology.* USDA Forest Serv. Tech. Rept. RM-32. pp. 167–170.

Gilbertson, D., M. Kent and F. Pyatt. 1985. *Practical Ecology for Geography and Biology.* London: Hutchinson and Co., Ltd.

Granger, O.E. 1983. The hydroclimatology of a developing tropical island: a water resources perspective. *Annals, Association of American Geographers* 73:183–205.

Greenland, D. 1979. *Modeling Air Pollution Potential for Mountainous Areas.* Occasional Paper #32. Institute of Arctic and Alpine Research. Boulder, CO: University of Colorado.

Greenland, D., and R. Yorty, 1985. The spatial distribution of particulate concentrations in the Denver metropolitan area. *Annals, Association of American Geographers* 75:69–82.

Hare, K. 1955. Dynamic and synoptic climatology. *Annals, Association of American Geographers* 45:152–162.

Harman, J. 1971. *Tropospheric Waves, Jet Streams, and United States Weather Patterns.* Commission on College Geography Resource Paper No. 11. Washington, D.C.: Association of American Geographers.

Harrison, J. and R. Larsen. 1977. Geography and planning: the need for an applied interface. *Professional Geographer* 29:139–147.

Hobbs, J. 1980. *Applied Climatology.* London: Dawson House.

Hubbard, K. and R. Hanks. 1983. Climate model for winter wheat yield simulation. *Journal of Climate and Applied Meteorology* 22:698–703.

Kalkstein, L. and P. Corrigan. 1986. A synoptic climatological approach for geographical analysis; assessment of sulfur dioxide concentrations. *Annals, Association of American Geographers* 76:381–395.

Kenzer, M.S. 1984. Comments from the outside: the sixth annual applied geography conference, 12–15 October, 1983. *Applied Geography* 4:85–86.

Kitada, T.K. Igarashi, and M. Owada. 1986. Numerical analysis of air pollution in a combined field of land/sea breeze and mountain/valley wind. *Journal of Climate and Applied Meteorology* 25:767–784.

Landsberg, H. and W. Jacobs. 1951. Applied climatology, in Malone, T. ed. *Compendium of Meteorology.* Boston: American Meteorological Society.

Marotz, G. 1983. Industry-government-academic cooperation: possible benefits for geography. *Professional Geographer* 35:407–415.

Mather, J. 1974. *Climatology: Fundamentals and Applications.* New York: McGraw-Hill Publishing Co.

Mather, J. 1978. *The Climatic Water Budget in Environmental Analysis.* Lexington, MA: D.C. Heath and Co.

Miller, D. 1987. History of Climatology. In J. Oliver and R. Fairbridge, pp. 338–346. *Encyclopedia of Climatology.* New York: Van Nostrand/Reinhold.

Muller, R. and A. Jackson. 1985. Estimates of climatic air quality potential at Shreveport, Louisiana. *Journal of Climate and Applied Meteorology* 24:293–301.

Muschett, F. 1981. Spatial distributions of urban atmospheric particulate concentrations. *Annals, Association of American Geographers* 71:552–565.

Olgyay, V. 1963. *Design with Climate.* Princeton, NJ: Princeton University Press.

Oliver, J. 1973. *Climate and Man's Environment: An Introduction to Applied Climatology.* New York: John Wiley and Sons.

Oliver, J. 1977. *Perspectives on Applied Physical Geography.* North Scituate, MA: Duxbury Press.

Oliver, J. 1981. *Climatology: Selected Applications.* New York: John Wiley and Sons.

Russell, J. 1983. Specialty fields of applied geographers. *Profession Geographer* 35:471–475.

Shelton, M. 1983. Basalts and groundwater management in the Deschutes river basin, Oregon. *Environmental Professional* 5:33–45.

Shelton, M. 1985. Modeling hydroclimatic processes in large watersheds. *Annals, Association of American Geographers* 75:185–202.

Smith, K. 1975. *Principles of Applied Climatology.* New York: McGraw-Hill.

Smith, K. 1986. Applied climatology. In Oliver, J. and R. Fairbridge eds., pp. 64–68. *Encyclopedia of Climatology.* New York: Van Nostrand/Reinhold.

Springer, E. 1972. *Foundations of Climatology.* San Francisco: W.H. Freeman, Inc.

Tarrant, J. 1984. Predicting USSR wheat production. *Applied Geography* 4:47–58.

Thomas, M. 1981. *The Nature and Scope of Climate Applications.* Canadian Climate Center Rep. 81–5. Downsview, Ontario: Atmospheric Environment Service.

Wilmott, C. 1982. On the optimization of the tilt and azimuth of flate-plate solar collectors. *Solar Energy* 28:205–216.
Winters, H. and M. Winters, eds. 1977. *Application of Geographic Research: Viewpoints from Michigan State University*. East Lansing, MI: Dept. of Geography.
Yoshino, M. 1975. *Climate in a Small Area*. Tokyo: University of Tokyo Press.

Glen A. Marotz
Department of Geography
Physics and Civil Engineering
University of Kansas
Lawrence, KS 66045
U.S.A.

8. Geomorphology: Praxis and Theory

Introduction

Geomorphology is the study of landforms, and applied geomorphology uses this science for practical benefit. In the last decade there has been a tremendous growth in the attention paid to applied geomorphology. This interest is manifested in the implementation of applied geomorphology courses and programs, and in a spate of recent publications. Indeed, a similar expansion in the applied aspects of all of geography has provided the impetus for this volume. Because applied geomorphology is the public face of our discipline, it is important to reflect on this visage, especially with regard to the discipline as a whole.

The number of recent articles and books published on applied geomorphology both surprised and perplexed me. Exploring some of the roots of the practical aspects of geomorphology has been a delightful learning experience; especially, I was surprised at how much has been written on this topic. This work has also given me a new respect for the breadth and substance of some of the applied work that has been done (e.g., Cooke 1984). However, I am also perplexed because the overall experience has left me with a curious feeling of 'much ado about nothing.'

This is partly because many articles published under the rubric of applied geomorphology make little or no contribution toward understanding of our environment, and they lend little to the advancement of geomorphological science. There is instead an emphasis on case studies and reviews of methods or concepts with attention focused upon data needs. This is not inherently bad; geomorphology must be made more accessible to managers, policy makers, and cooperative researchers from other disciplines (Craig and Craft 1982b). Applied geomorphology publications serve this purpose through the generalization of the science and by using case studies to illustrate particular methods and appropriate theories. This broadens the prospect of additional applications. Unfortunately this work is circulated mainly among geomorphologists rather than managers; thus, the exposure to potential users is reduced. Furthermore, few of these works represent state-of-the-art geomorphology and are frequently the otherwise unpublishable summaries or overviews of contract work. Thus, our bookshelves experience an increase in weight, but not substance. The core of our field is not enriched.

Another concern arises from the distinction of 'applied' from 'pure' or 'scientific' geomorphology. Certainly the term 'applied geomorphology' has meaning for us, implying something different than mainstream geomorphology. The many facets of the term invoke images of contract work, research funding, improving the human condition, disciplinary relevance, or perhaps personal gratification. It is the extension of our science beyond the walls of academe. Jones

M. S. Kenzer (ed.), Applied Geography: Issues, Questions, and Concerns, 115–131.
© 1989 *Kluwer Academic Publishers.*

(1980) has called the rapid integration of applied geomorphology into the discipline one of the most fundamental developments in the field. The purpose of this essay is to examine the definition, development, and structure of applied geomorphology, and to discuss its potential contributions to the rest of the discipline.

It is both necessary and important that geomorphology make direct contributions to the common welfare. If we perform good works with contemporary relevance, then our science will gain in stature. Applied work also provides the chance to use our theories and methods in the real world, the ultimate proving ground for scientific geomorphology.

Kuhn (1970, p. 10), in discussing 'normal science,' defines his central term of paradigms as illustrated by 'examples of actual scientific practice – examples which include law, theory, application, and instrumentation together,' and he suggests that these examples will 'provide models from which spring particular coherent traditions of scientific research.' If we accept this view, then the application of geomorphological science is an integral part of its existence; without application, it is not a 'normal science.' It is also corollary that the repeated testing of our present tenets of geomorphology through rigorous usage should improve our science through either paradigm refinement or paradigm shifts (Kuhn 1970). Applied geomorphology is, therefore, not an epiphenomenon. It is certainly inarguable that major advances in geomorphology have sprung directly from practical research. But, it is also necessary to evaluate the relative roles of applied and pure geomorphology today.

Defining Applied Geomorphology

There are surprisingly few formal definitions of applied geomorphology. Goudie (1985, p. 23) uses an essay by Hails (1977b) to define applied geomorphology as:

> The application of geomorphology to the solution of miscellaneous problems, especially to the development of resources and the diminution of hazards.

Similarly, the definition by Brunsden *et al.* (1978, p. 251) is:

> ...the application of geomorphological techniques and analysis to the solution of planning, environmental management, engineering or similar problems.

These definitions are unambiguous, but somewhat general. Nevertheless we understand the broad nature of the potential applications of geomorphology. In the past, however, the notion of applied geomorphology has also taken a number of other forms, some of which are quite narrow. In geography, for example, the application of geomorphology has often meant the furnishing of baseline information for cultural geographers. It is most disconcerting to hear this role advocated by geomorphologists themselves. Russell (1949, p. 10), as president of the Association of American Geographers, called for the admission of geographical geomorphologists who practice a science that 'really tells us what is present in a landscape and tells us exactly where each form is to be found,' as a basis for regional cultural geography. The same notion is reflected in the views of both Bryan (1950) and Kesseli (1950). Bryan (1944, p. 188), addressing the work of the geomorphologist, wrote that 'The interrelationships of his work with "human" geography will be close, and he should work with full knowledge and understanding of the use to which his product may be put.' Certainly we still recognize the value of cooperative exchange between human and physical geography, including geomorphology. However, there is also an obligation to recognize the discrete contributions that geomorphology can make in its own right, a point well argued by Orme (1985).

Steers (1971) defines applied geomorphology as the use of the methods and concepts of geomorphology to particular places, rather than practical utilization. Coates (1971, p. 6) defined an environmental geomorphology that is clearly linked to applied geomorphology as addressed by Hails (1977b): 'Environmental geomorphology is the practical use of geomorphology for the solution of problems where man wishes to transform landforms or to use and change superficial processes.'

Most other authors and editors, however, presume the reader understands what applied geomorphology is and do not define the term explicitly (e.g., Cooke and Doornkamp 1974; Craig and Kraft 1982b; Costa and Fleisher 1984; or Hart 1986); even Hails (1977b) avoided the stricture of a refined statement. Brunsden (1985, p. 225), for example, avoided strict definition of applied geomorphology and described its aims instead:

...to assist in the efficient discovery, assessment and wise management of the earth's finite resources, to prevent environmental deterioration and to avoid or prevent natural hazards.

But, for the most part, we are left to work through the histories and benefits and examples of applied geomorphology and then form our own opinions. In many circumstances this experiential definition would be suitable (given a common nexus), and if we were to consider only recent applied geomorphology publications this would not be a problem. They are almost uniformly aimed at representing the practical aspects of geomorphological science as benefits mankind. Still, Jones (1980, p. 50) found it worthwhile to caution that 'it is essential that the nature of applied geomorphology be clearly defined, for the term has come to be loosely associated with a very wide range of activities.'

We shall consider applied geomorphology in the broadest, modern sense, essentially as used by Hails (1977b) and defined by Brunsden (1985). This interpretation includes using geomorphological theories and methods to address environmental problems, in any capacity. This approach is adopted because these definitions afford the greatest scope for practicing geomorphology in the real world. By accepting these definitions we can also reinforce the notion that application is an integral part of the practice of normal science.

The Soviets were perhaps first to formalize a relationship between the science and practice of geomorphology. Indeed, according to the needs and dictates of the state, there is no basis for science without utilization for the common good. Therefore, by definition, all geomorphology is aimed at application, as succinctly expressed by Shchukin (1960, p. 36):

The most complete definition of geomorphology as a science and the one that corresponds best to the dialectical method of cognition is the following: geomorphology is concerned with the study of the relief of the earth's surface for the purpose of its practical utilization in the economic activities of man.

And, with regard to the role of geographers in government, Sochava (1970, p. 731) writes that the practitioner must be prepared 'to advise on problems relating to ... the protection of landscapes against undesirable transformations.' From this viewpoint, there is no independent rationale for 'pure' geomorphology: no reason exists to adopt scientific concern over matters lacking social consequences. Thus, there is no need to be concerned with applied geomorphology *per se*, because the expression is redundant, a view in harmony with that of by Craig and Craft (1982b, p. v):

> Geomorphologists tend to consider all geomorphologic research to be applied. In the sense that each advance in knowledge provides a clearer view of how the earth works, we must all be applied scientists.

Both views bring us back to the earlier proposition that if geomorphology is, or is to be, a normal science, then application must be a foundation of the field.

The Development of Applied Geomorphology

Why is applied geomorphology receiving so much attention at present? This question is especially pertinent given the tradition of practically-based geomorphological research. There are five possible reasons: the relative immaturity of the field and the meager history of applied studies; the advent of process geomorphology; the environmental movement of the 1960s; the growth of urban geomorphology; and recognition of the importance of humans as geomorphological agents.

Geomorphology as a Young Science

First, as an independent discipline, geomorphology is young relative to most sciences, the number of practitioners is small, and many ideas still enter the field from other branches of knowledge, especially geology, engineering, climatology, oceanography, forestry, and hydrology. Many of the 'classic' contributions to geomorphology are immigrants from other fields, now accorded citizenship. Nevertheless, the publication explosion in applied geomorphology over the last two decades dwarfs these earlier efforts and minimizes this tradition. To ignore the earlier years, however, is to overlook the gestation period for many of the specialty fields of geomorphology, especially fluvial, coastal, and periglacial, and the applied aspects of geomorphology in general. Major streams of this history are reviewed below.

The early works of European engineers, like Guglielmini (1697, cited in Kesseli 1941), Surell (1841), and Dause (1857) are particularly noteworthy, especially with regard to the understanding of processes. Their work on fluvial processes resulted from practical interests in stream control, and represent the foundation of modern concepts of geomorphic equilibrium (Kesseli 1941; Dury 1966). Tinkler (1985) pushes this time scale back even further, showing the basis of process geomorphology in the work of early civil engineers, such as da Vinci.

If we look to the geographic and geologic roots of geomorphology, we can also discern a long heritage. According to Hartshorne (1958, p. 99):

> Prior to the eighteenth century few students of geography felt any need to determine the status of their subject in the general field of knowledge; its importance was sufficiently assured by popular interest and general utility. In that century, however, an increasing number of students became concerned to establish geography as an integral field of knowledge, rather than merely a utility service of commerce and government...

Recall that physical geography (including geomorphology) was a primary component of the early geography Hartshorne addressed. And, from a geological background, Davies (1969) suggested that the Industrial Revolution brought British geology out of the ivory towers and into the trenches. The work of these 'mechanic' geologists (Davies 1969, p. 139) included background research for the design and construction of canals and roads; thus, applied geomorphology (although in a different guise) was not uncommon in the late eighteenth century.

We can likewise trace the blossoming of the modern era in North American geomorphology directly back to commissioned landform studies by nineteenth-century geologists. The greatest boon resulted from the U.S. government expeditions to the American West. Primarily practical, especially in the early explorations, this work was colorfully described by the first director of the U.S. Geological Survey, C. King (1880, p. 4):

> Eighteen hundred and sixty-seven, therefore, marks, in the history of national geological work, a turning point, when the science ceased to be dragged in the dust of rapid exploration and took a commanding position in the professional work of the country.

Dutton (1889), Gilbert (1877, 1890), and Powell (1875) all made substantial contributions to modern geomorphology as outgrowths of their duties with the western surveys. Gilbert (e.g., 1877, 1890, 1914, 1917), in particular, laid the backbone for process geomorphology as a result of his applied work. It is not obvious whether these men thought they were practicing pure or applied geomorphology. That science and its use were coincident was probably taken for granted.

For most of the early twentieth century geomorphology withered. This may be in part a reaction to the excesses of Davis (Beckensale 1976), a response to extreme environmental determinism, or to Hartshorne's rejection of physical geography (Sauer 1941). Davis also attributed the atrophe to deficiencies in the teaching system. In describing the swing from physiography to commercial geography, he gave three reasons for this decline, one of which is pertinent here (Davis 1932, p. 229): 'they [physiography teachers] *linked it up so little with its human consequences* that their pupils did not think that it had any' (emphasis added). This, despite the admonition of Fenneman (1916, p. 25) that in the description of physiographic regions 'It is ... a matter of coordinate importance that these divisions shall also be useful in the consideration of the effect of topography on human affairs.'

For whatever reason, what we today consider applied geomorphology, and indeed geomorphology in general, was relatively stagnant for about fifty years. A rejuvenation can be attributed largely to Strahler (1952) and the search for process oriented explanation that contributed to the more intensely scientific training most geomorphologists presently receive. In his seminal paper, Strahler (1952, p. 923) explicitly claimed his theoretical work was an outgrowth of contract work for the Office of Naval Research, Geography Branch. There is also a substantial body of specifically applied work representing the 1960s and 1970s (briefly reviewed by Cooke and Doornkamp, 1974; and exemplified by Coates 1973), much of which explicitly relates landforms and cultural activities (Zakrzewska 1967).

Process Geomorphology

A second reason for the recent explosion in applied geomorphology has been tied to the development of modern process geomorphology, an idea espoused directly by Hart (1986, p. 129):

To a large extent, applied studies and process studies go together, because applied studies rely on there being an explanation for geomorphological phenomena, and explanation is impossible without understanding process.

There were few geomorphologists actively seeking detailed, process oriented explanations for landform development before the 1960s. *Ipso facto* applied geomorphology could not have prospered before then.

The rationale for this supposition is that only through an understanding of geomorphological processes can we begin to predict landform behavior, a necessary element in environmental management. This theme is found in Cooke and Doornkamp (1974), Jones (1980), Matthewson and Cole (1982), Costa (1984), and Nordstrom (1986), *inter alia* (although not in the context presented here). My argument presupposes, therefore, that those whom we might call traditional geomorphologists had little to offer to the welfare of society. This is clearly specious logic, as demonstrated by the long practice of applied geomorphology.

Environmentalism

A third line of reasoning claims that applied geomorphology prospered as an outgrowth of the 'ecological movement' of the 1960s and 1970s. From a general academic desire to demonstrate relevance, many geomorphologists began to consider how they could contribute to the management of environmental concerns. This was a primary impetus for the growth of environmental geomorphology (e.g., Cooke and Doornkamp 1974; Coates 1971, 1973), and it is a frequent message (e.g., Butler 1971; Everett 1982). As we have seen above, the definition of environmental geomorphology is synonymous with what we today define as applied geomorphology.

Urban Geomorphology

A fourth reason for the present development of applied geomorphology has been the recognition of the value of integrating basic geomorphological principles into the study of urban environments. We see here, in particular, a strong tie between planning and public administration, engineering, geology, geography, and geomorphology; this tie is strongly evidenced by the selection of papers presented in Coates's (1976) *Geomorphology and Engineering*. Most of the work concerning urban geomorphology traditionally has been reported in the 'gray literature' of technical reports and monographs. However, the appearance of useful overviews of contributions from geomorphology (e.g., Detwyler and Marcus 1972; Legget 1973; Cooke 1976; Utgard et al. 1978) have spurred greater awareness of the value of geomorphological research efforts in this field.

Some examples of applied urban geomorphology include the hazards research of Ives et al. (1976) on avalanches, Bryan and Price (1980) on cliff erosion, Cooke (1984) on landslides and flooding, Gupta (1984) on flooding, subsidence, and sedimentation, and sections in Bird and Schwartz (1985) on coastal erosion. Other examples with strong implications for planners include the works of Walling and Gregory (1970), Cooke and Jones (1980), Cooke et al. (1982), and an interesting contribution by nongeomorphologists, Arlinghaus and Nystuen (1987) on urban slope gradients and bus routes.

Anthropogeomorphology

Finally, there is a thin but rich tradition of anthropogeomorphological research with strong implicit or explicit ties with applied geomorphology (Walker and Mossa 1986). Much of the work describing the role of humans as geomorphological agents is commonly referenced in applied geomorphology literature. Whenever the results of human activities impact the physical landscape, and if these impacts are documented, this record can then be applied toward future endeavors with the aim of avoiding, reproducing, or improving possible outcomes (Bunge 1973).

Some primary contributions to this approach are the writings of Fischer (1915) and Sherlock (1922), two of the first broad commentaries on the subject, and Marsh (1864) and Woeikof (1901), concerning increased erosion from devegetation (among other topics). Several of the chapters in Thomas (1956) address anthropogeomorphology directly and are in an applied spirit. Golomb and Eder (1964) proposed the new science of anthropogeomorphology – of course it is neither new nor a science in its own right – and Brown (1970, p. 75) introduced the concept of 'Man, the geomorphological process' in his wide-ranging address. He also tied his remarks to the conservation movement, thus touching on two of the subject areas mentioned above. More recent discourse on anthropogeomorphology can be found in Goudie (1982) and Nir (1983).

The Structure of Applied Geomorphology

There have been several efforts to characterize the structure of applied geomorphology. Some aim at describing the applications in terms of their subject matter; others examine the nature of the practice. Verstappen (1983) lists five main fields for geomorphological applications: earth sciences, including geomorphological mapping; environmental studies, especially with regard to natural hazards; rural development and planning; urban development and planning; and engineering, especially assessment of engineering works. Goudie (1985, p. 24) recognizes six similar categories in the organization of Cooke and Doornkamp (1974).

This type of categorization allows us to get a simple overview of what people are doing, but it tells us very little about the actual practice of geomorphology. Classifications are, in fact, just a broad means of identifying what applied geomorphology is. In evaluating applied geomorphology it is more valuable to examine how geomorphologists ply their trade.

Applied geomorphology can be practiced at a number of levels, which can be arranged into a hierarchy of complexity based upon requirements of responsibility and expertise. At the simplest level is (1) the retrieval of archival information (including map data). In order of increasing complexity, the subsequent levels are (2) basic, directed data acquisition, (3) data analysis and interpretation, (4) experimental design for specific or general information requirements, (5) the design and implementation of complex, transdisciplinary research projects, and (6) the participation in basic public policy formation at superdisciplinary levels.

Archival Research

Archival research is at the lowest level of the hierarchy for several reasons. First, it can be accomplished with a minimum of geomorphological training, and it is probably best practiced by research librarians armed with a list of key words and concepts. Further, the mere accumulation of second-hand data cannot reveal new insights concerning the geomorphic environment. Finally, this level of activity requires that the research is directed from a higher level.

As it stands, archival research alone does not contribute to our understanding of the environment, and it is not science. The formation of data inventories from secondary sources alone is also not really applied geomorphology. However, both activities may contribute substantially to a project's success when the results are integrated and analyzed as part of a research program (see Fisher 1984; Gares and Sherman 1985, *inter alia*).

Directed Data Acquisition

Directed data acquisition, the next level in the hierarchy, requires a broader based understanding of the specific methods of geomorphology, although little or no scientific expertise is needed. Directed data acquisition involves the technical aspects of a geomorphological project and implies that pertinent parameters and data needs are decided at a higher level. The practitioner has little discretion in experiment design.

The tasks associated with typical data acquisition require some level of expertise with sampling methods and particular techniques. As there are generally economic considerations associated with the implemention of most projects, the number of necesary samples, and often the manner in which they are gathered, are frequently dictated by a project manager. This step is often crucial because the data form the basis of later analysis (e.g., Everett 1982). Nevertheless, this stage is not a primary application of the scientific skills of geomorphology, and the work is usually done by field or research assistants who might not require extensive training.

Data Analysis and Interpretation

In geomorphology, the roles of data analysis and interpretation represent a quantum leap over the duties described above. First, substantial expertise is required to recognize appropriate techniques to reduce and analyze data. The choice of inappropriate analytic methods (e.g., parametric statistics for non-normally distributed data) can lead to spurious results and a misapprehension of environmental relationships. Also, the analyst will generally have detailed knowledge of the methods used to obtain the data and may even have set the sampling procedures.

Interpretation involves knowledge of the discipline beyond technique. Here, for the first time, science becomes a major requisite of successful action. The interpretation and explanation of results requires an understanding of the fundamental relationships governing the phenomenon under investigation. This distinction between acquisition and analysis, relative to interpretation, is as fundamental as the difference between measuring and describing stream velocities and energy gradients, and using those data to explain the relationship between the two variables. A statistician can recognize the statistical explanation of stream velocity by energy gradient. The geomorphologist can understand and explain the physical relationship. This is the difference between a technician and a scientist. This type of work is discussed in Jones (1984), Ollier (1977), and Knox (1987).

Experimental Design

The prerequisites of successful experimental design include substantial understanding of the experiment's objectives (why are we doing this?), the physical processes and responses in the geomorphological problem (what is going on?), and the appropriate means of measuring and analyzing the pertinent parameters (how can we do this?). An understanding of appropriate theory is important in successful experimental design, both from a scientific and practical aspect. At this level interaction between the scientist and the manager becomes particularly important to achieve project goals. Note that manager is used as a generic term: the user of the applied geomorphology.

Careful communication between managers and scientists is important to identify which part of an environmental interaction is relevant to a particular project. At this level, the scientist must rely on management to have recognized and circumscribed the problem. In the case of funded research, a contracting agency may dictate explicitly the nature of the work, and project supervision will likely occur through communication at this level. It is the geomorphologist's obligation to pre-appraise the validity of project specifications and suggest any modifications that might enhance the experiment's viability. This allows the project to be operationalized within the bounds of accepted geomorphological practice (e.g., using the process-response model of Krumbein 1963). Within these constraints the practitioner must then design the experiment to ensure that all pertinent data are obtained in a usable form, see that the appropriate analyses are performed, and be responsible for the interpretation and presentation of the results to the sponsors.

The importance of communication cannot be over stressed. Sherman and Gares (1982) illustrate how close interaction between planners and scientists can lead to overall benefits by setting realistic criteria and streamlining both information requirements and the decision making process. Brunsden (1985) similarly emphasizes the value of early and continuing communication between geomorphologists and clients. This cooperative effort will directly address White's lament (1966, p. 427) over 'the gap between the technology that is known and the technology that is applied.' Although he was referring to problems of water use in arid lands, the situation is common to most environmental issues.

Failure of the project manager to define the problem adequately may lead to invalid experimentation, measurement of irrelevant parameters, and analysis of extraneous information. The manager is obligated to set specific project objectives, and it is the responsibility of the geomorphologist to provide the information necessary for rational decision making. Interaction at both levels optimizes the chances for project success and enhances the economies of basic experimentation.

Project Design and Implementation

The design and implementation of complex projects requires familiarity with several disciplines at both scientific and managerial levels. Here the geomorphologist transcends his or her discipline by directing and perhaps administering a project where earth surface processes are only one aspect of a problem. This frequently requires an appraisal of the importance of geomorphic processes with respect to other concurrent ecosystem processes (e.g., biological, economic, or chemical processes). It also implies that the geomorphologist, as scientist, assumes responsibility for the synthesis of information deriving from the different parts of the investigation. As manager, the geomorphologist is responsible for assessing the planning alternatives and suggesting appropriate decisions. Where the geomorphological aspect of the problem is central to the project, the applied geomorphologist plays primary roles as both scientist and manager.

At this point we must distinguish between practice and practitioner. As we consider the upper levels of the hierarchy, the obligations of the applied geomorphologist require increasing managerial skills; many of the specialist characteristics give way to the generalist talents of a successful manager. Understanding the basic principles of the discipline is still essential for the satisfactory exercise of responsibilities.

It is not conceit to claim these roles for geomorphologists and it is not an exclusive claim. Many authors have discussed the increasing complexity of environmental problems and a concomitant need for more interdisciplinary work between geomorphologists and others (Butzer 1973; Craig and Craft 1982b; Brunsden 1985; *inter alia*). It is clear that someone or some group will have to accept responsibility for guiding any project.

Public Policy Formation

At the top of this hierarchy is the applied geomorphologist taking an active role in public policy formation. This last step is taken infrequently because it is distant from the discipline's core, there are few opportunities, the performance pressures are extreme, and because training in the needed skills is usually not part of the geomorphologist's repertoire. This is an important theatre of operation, nevertheless.

In the realm of policy formation the geomorphologist serves the discipline in two main capacities. First, using geomorphic considerations as a basis for planning increases the visibility of the subject and will demonstrate the value of understanding earth surface processes and landform responses. Brunsden (1985) offers several examples of the rewards of this exposure. Second, as champions of this vision, the geomorphologist making policy has the opportunity to direct applications at other positions in this hierarchy, thereby promoting the discipline's viability by addressing real world problems. In a sense, this is the 'motherhood' role that protects geomorphology in the public arena and helps it to prosper. Coates (1984), reviewing the role of the geomorphologist in public policy, demonstrates precisely the opportunities available for responsible action. Nordstrom and Renwick (1984), Williams and Sothern (1986), and Nordstrom (1986) are examples of attempts to incorporate geomorphological criteria into public policy designed especially to control development in areas of particular natural interest or hazard.

The General Structure

Reviewing the practice of applied geomorphology at each level of this hierarchy, several generalizations can be drawn. First, little or no geomorphological expertise is required to perform well at the two lowest levels. Even considering direct data acquisition, one can argue that the purely technical skills required can be taught apart from the particular training of the geomorphologist. Scientific skills are not required where appropriate techniques can be accomplished with 'cookbook' approaches. This is not to denigrate these activities, as they can be both rewarding and productive (Brunsden 1985).

At intermediate levels, where most applied geomorphology is practiced, the emphasis is almost entirely on the scientific aspects of geomorphology. Although the needs of a particular project may dictate the rationale and the appropriate parameters, the subsequent experimental design and implementation should proceed essentially independent of managerial goals. In these stages it is difficult to ascribe any attributes to applied geomorphology that would not also be fitting for 'pure' geomorphology. The application controls the subject and the use of the findings; it should not control the methodology.

At the most complex levels of applied geomorphology the emphasis shifts from the discipline to the disciple. One need not be adept at field experimentation or theory to be a successful actor in the public policy domain. Although skills at all levels can contribute to performance at these stages, they are not central to the role of policy formation. Instead, the most effective applications for the benefit of geomorphology might spring from the works of the generalist who can ably recognize where the discipline can provide new opportunities for resource management, offer alternatives for present problems, and perform well in the boardroom.

Conclusions

Gerrard (1984, p. v) claimed that 'applied geomorphology is essentially a recent phenomenon.' This is a naïve and somewhat self-serving perspective, and it is unrealistic, as shown above (also see Hart 1986). Furthermore, there should be neither methodological nor theoretical distinctions between applied and 'pure' geomorphology. The primary differences involve rationales for research, sponsorship, and perhaps the generality of the results (hence the abundance of case studies in applied geomorphology). However, there is not a one-to-one correspondence between applied and scientific geomorphology throughout the hierarchy outlined above. This is not troublesome, as the important contributions of our discipline to public welfare come not only through the rigors of scientific exploration, but also by forging recognized links between science and politics. If we limit our participation in the policy decision-making process to the provision of information to others, and if we eschew an active role in policy analysis and formulation, then we forfeit the right to argue the relevance of our discipline. Consequently, we will be supplying geomorphology to others, rather than applying it ourselves. We will then have to rely on benevolent 'others.'

It is also important to note that this is not an argument for all geomorphologists to become politicians, because each should operate at the most comfortable level. This is instead a reminder that there are higher purposes that we can strive toward. Kesseli (1950) argued that we are naive if we expect other scientists to do the hard work while geomorphologists generalize and reap the rewards. Unfortunately, he also advocated that geographical geomorphologists back-off and merely describe landforms independent of genesis or process. The basis of this approach is faulty, however.

In sponsored research, particularly contract work, scientists turn their findings over to the commissioning agency. These efforts are then appraised on their ability to meet specific management criteria. A problem frequently encountered in this process is the difficulty of communication between the scientist and the manager; the latter tries to make informed decisions in situations where there is often no optimal solution. When scientists participate actively in decision making, the selection of choices is facilitated because the information flow to managers is enhanced. If we presume the importance of the role of science in the policy sphere, then there is every reason to claim an appropriate position for geomorphologists.

It is also fundamental that geomorphologists retain objectivity in performing appointed tasks. This can become increasingly difficult as the level of interaction rises, but it remains basic to science (Johnston 1981; Nordstrom 1983). The avoidance of bias is also one reason why logical positivism will remain a keystone of most applied research. It is also crucial, more difficult still, to remain objective about the nature of the work that we accept. Where research arenas are dictated by funding sources, there is a risk that the direction of science will be subordinated to social, political, or commercial expediency. Chorley (1978, p. 11) pointed this out, warning that:

> ...utilitarian approaches to geomorphology will result either in large-scale work of which the intellectually sterile taxonomic morphological mapping is the most depressing precursor, or in a piecemeal concentration on small-scale realist systems ... [and] the quest for utility not only conditions the objects of our work, but also the manner in which theory may emerge ...

The prescience of this admonition is apparent in surveys of applied geomorphology. However, it will always be difficult to balance the search for disciplinary relevance, sources of funding, and some ethereal notion of the good of our science. But, our independence, and perhaps our identity, is at stake. We must heed Harvey's caution (1984, p. 7) about applied geography in general:

> Geography ... is far too important to be left to generals, politicians, and corporate chiefs. Notions of 'applied' and 'relevant' geography pose questions of objectives and interests served. ... There is more to geography than the production of knowledge and personnel to be sold as commodities to the highest bidder.

We can substitute geomorphology into his quotation and recognize the warning. Geomorphologists must take responsibility for directing the discipline. If our science is good, then applications will follow as part and parcel of normal science. If this is an acceptable notion then we are each freed to pursue an understanding of the physical landscape. The opportunity to do applied work will broaden our backgrounds and open new avenues for research. In the application of our science we can be the users rather than the used.

References

Arlinghaus, S.L. and Nystuen, J.D. 1987. Geography of city terrain based on bus routes. *Geographical Review* 77:183–195.
Beckensale, R.P. 1976. The international influence of William Morris Davis. *Geographical Review* 66:448–466.
Bird, E.C.F. and Schwartz, M.L. eds. 1985. *The World's Coast line*. Van Nostrand Reinhold: Stroudsburg, PA.
Brown, E.C. 1970. Man shapes the earth. *Geographical Journal* 136:74–84.
Brunsden, D. 1985. Geomorphology in the service of society. in R.J. Johnston ed., pp. 225–257. *The Future of Geography*. Methuen: London.
Brunsden, D., Doornkamp, J.C. and Jones, D.K.C. 1978. Applied geomorphology: a British view.' in C. Embleton, D. Brunsden and D.K.C. Jones eds., pp. 251–262. *Geomorphology: Present Problems and Future Prospects*. Oxford University Press: Oxford.
Bryan, K. 1944. Physical geography in the training of the geographer. *Annals of the Association of American Geographers* 34:183–189.
Bryan, K. 1950. The place of geomorphology in the geographic sciences. *Annals of the Association of American Geographers* 40:196–208.
Bryan, R.B. and Price, A.G. 1980. Recession of the Scarborough Bluffs, Ontario, Canada. *Zeitschrift für Geomorphologie*. Supplement Band 34:48–62.
Bunge, W.W. 1973. The geography of human survival. *Annals of the Association of American Geographers* 63:275–295.
Butler, J.H. 1971. Geomorphology and decision-making in water resource engineering. in D.R. Coates ed., pp. 81–89. *Environmental Geomorphology*. State University of New York: Binghamton.
Butzer, K.W. 1973 Pluralism in geomorphology. *Proceedings of the Association of American Geographers* 5:39–43.
Chorley, R.J. 1978. Bases for theory in geomorphology. in C. Embleton, D. Brunsden, and D.K.C. Jones eds., pp. 1–13. *Geomorphology: Present Problems and Future Prospects*. Oxford University Press: Oxford.
Coates, D.R. 1971. Introduction to environmental geomorphology. In D.R. Coates ed., pp. 5–6. *Environmental Geomorphology*. State University of New York: Binghamton.
Coates, D.R. ed. 1973. *Environmental Geomorphology and Landscape Conservation*. Dowden, Hutchinson & Ross: Stroudsburg, PA.
Coates, D.R. ed. 1976. *Geomorphology and Engineering*. Dowden, Hutchinson & Ross: Stroudsburg, PA.
Cooke, R.U. 1976. Urban geomorphology. *Geographical Journal* 142:59–65.
Cooke, R.U. 1984. *Geomorphological Hazards in Los Angeles*. George Allen & Unwin: London.
Cooke, R.U., Brunsden, D., Doornkamp, J.C. and Jones, D.K.C. 1982. *Urban Geomorphology in Drylands*. United Nations University and Oxford University Press: Oxford.
Cooke, R.U. and Doornkamp, J.C. 1974. *Geomorphology in Environmental Management*. Clarendon Press: Oxford.
Cooke, R.U. and Jones, D.K.C. 1980. Suitable sites for cities. *Geographical Magazine* 52:356–358.
Costa, J.E. 1984. Physical geomorphology of debris flows. In J.E. Costa and P.J. Fleisher, pp. 268–317.
Costa, J.E. and Fleisher, P.J. eds. 1984. *Developments and Applications of Geomorphology*. Springer-Verlag: Berlin.
Craig, R.G. and Craft, J.L. 1982a. *Applied Geomorphology*. George Allen & Unwin: London.

Craig, R.G. and Craft, J.L. 1982b. Preface. In R.G. Craig and J.L. Craft eds. 1982a, pp. v-vi.

Dausse, M.F.B. 1857. Note sur un principe important et nouveau d'hydraulique. *Comptes Rendus Académie Scientifique* (Paris) 44:757–766.

Davies, G.L. 1969. *The Earth in Decay.* American Elsevier Publishing Company: New York.

Davis, W.M. 1932. A retrospect of geography. *Annals of the Association of American Geographers* 22:211–230.

Detwyler, T.R. and Marcus, M.G. 1972. *Urbanization and Environment.* Duxbury Press: Belmont, CA.

Dury, G.H. 1966. The concept of grade. In G.H. Dury ed., pp. 211–233. *Essays in Geomorphology.* Heinemann Educational Books: London.

Dutton, C.E. 1889. On some of the greater problems of physical geology. *Bulletin of the Philosophical Society of Washington* 11:51–64.

Everett, A.G. 1982. Geomorphic process data needs for environmental management in R.G. Craig and J.L. Craft, pp. 1–16.

Fenneman, N.M. 1916. Physiographic divisions of the United States. *Annals of the Association of American Geographers* 6:19–98.

Fischer, E. 1915. Der Mensch als geologischer Faktor. *Zeitschrift Deutsche Geologische Gesellschaft* 76:106–148.

Fisher, J.J. 1984. Regional long-term and localized short-term coastal environmental geomorphology inventories. In J.E. Costa and P.J. Fleisher, pp. 68–96.

Gares, P.A. and Sherman, D.J. 1985. Protecting an eroding shoreline: the evolution of management response. *Applied Geography* 5:55–69.

Gerrard, J. 1984. Introduction. *Zeitschrift für Geomorphologie,* Supplement Band 51:v-viii.

Gilbert, G.K. 1877. *Report on the Geology of the Henry Mountains.* U.S. Geographical and Geological Survey of the Rocky Mountain Region: Washington, D.C.

Gilbert, G.K. 1890. *Lake Bonneville.* U.S. Geological Survey Monograph Number 1: Washington, D.C.

Gilbert, G.K. 1914. *The Transportation of Debris by Running Water.* U.S. Geological Survey Professional Paper Number 86: Washington, D.C.

Gilbert, G.K. 1917. *Hydraulic Mining Debris in the Sierra Nevada.* U.S. Geological Survey Professional Paper Number 105: Washington, D.C.

Golomb, B. and Eder, H.M. 1964. Landforms made by man. *Landscape* 14:4–7.

Goudie, A. 1982. *The Human Impact: Man's Role in Environmental Change.* MIT Press: Cambridge, MA.

Goudie, A. ed. 1985. *The Encyclopedic Dictionary of Physical Geography.* Basil Blackwell: New York.

Gupta, A. 1984. Urban hydrology and sedimentation in the humid tropics. In J.E. Costa and P.J. Fleisher, pp. 240–267.

Hails, J.R. ed. 1977a. *Applied Geomorphology.* Elsevier Scientific Publishing: Amsterdam.

Hails, J.R. ed. 1977b. Applied geomorphology in perspective. In J.R. Hails ed. 1977a, pp. 1–8.

Hart, M.G. 1986. *Geomorphology: Pure and Applied.* Allen and Unwin: London.

Hartshorne, R. 1958. The concept of geography as a science of space, from Kant and Humboldt to Hettner. *Annals of the Association of American Geographers* 48:97–108.

Harvey, D. 1984. On the historical and present condition of geography: an historical materialist manifesto. *Professional Geographer* 36:1–11.

Ives, J.D., Mears, A.I., Carrara, P.E. and Bovis, M.J. 1976. Natural hazards in mountain Colorado. *Annals of the Association of American Geographers* 66:129–144.

Johnston, R.J. 1981. Applied geography, quantitative analysis and ideology. *Applied Geography* 1:213–219.

Jones, D.K.C. 1980. British applied geomorphology. *Zeitschrift für Geomorphologie.* Supplement Band 36:48–73.

Jones, J.R. 1984. Computer applications in coastal geomorphology. In J.E. Costa and P.J. Fleisher 1984, pp. 38–67.

Kesseli, J.E. 1941. The concept of the graded river. *Journal of Geology* 49:561–588.

Kesseli, J.E. 1950. Geomorphic landscapes. *Yearbook of the Association of Pacific Coast Geographers* 12:3–10.

King, C. 1880. *First Annual Report of the United States Geological Survey.* Government Printing Office: Washington, D.C.

Kniffen, F.B. 1973. Richard Joel Russell, 1895–1971. *Annals of the Association of American Geographers* 63:241–249.

Knox, J.C. 1987. Historical valley floor sedimentation in the upper Mississippi valley. *Annals of the Association of American Geographers* 77:224–244.

Kuhn, T.S. 1970. *The Structure of Scientific Revolutions.* 2nd Edition. University of Chicago Press: Chicago.

Krumbein, W. 1963. A geological process-response model for the analysis of beach phenomena. *Annual Bulletin of the Beach Erosion Board* 17:1–15.

Leggett, R.F. 1973. *Cities and Geology,* McGraw-Hill Publishing: New York.

Marsh, G.P. 1864. *Man and Nature, or Physical Geography as Modified by Human Action,* Scribners: New York.

Mathewson, C.C. and Cole, W.F. 1982. Geomorphic processes and land use planning, South Texas barrier islands. In R.G. Craig and J.L. Craft, 1982a, pp. 131–147.

Nir, D. 1983. *Man, a Geomorphological Agent,* D. Reidel Publishing: Dordrecht.

Nordstrom, K.F. 1983. 'Science' and the funding of physical geographers. *Professional Geographer* 35:469–470.

Nordstrom, K.F. 1986. Beach conservation and enhancement: the basis for a national policy on coastal erosion in the United States. *Journal of Shoreline Management* 2:13–34.

Nordstrom, K.F. and Renwick, W.H. 1984. A coastal cliff management district for protection of eroding high relief coasts. *Environmental Management* 8:197–203.

Ollier, C.D. 1977. Terrain classification: Methods, applications and principles. In J.R. Hails 1977a, pp. 277–316.

Orme, A.R. 1985. Understanding and predicting the physical world. In R.J. Johnston ed., pp. 258–275. *The Future of Geography.* Methuen: London.

Powell, J.W. 1875. *Exploration of the Colorado River of the West, and its Tributaries Explored in 1869, 1870, 1871, and 1872.* Smithsonian Institution: Washington, D.C.

Russell, R.J. 1949. Geographical geomorphology. *Annals of the Association of American Geographers* 39:1–11.

Sauer, C.O. 1941. Foreword to historical geography. *Annals of the Association of American Geographers* 31:1–24.

Schukin, I.S. 1960. The place of geomorphology in the system of natural science and its relationship with integrated physical geography. *Soviet Geography* 1:35–43.

Sherlock, R.L. 1922. *Man as a Geological Agent.* Witherby: London.

Sherman, D.J. and Gares, P.A. 1982. Environmental Strategies: a case study of systematic evaluation. *Environmental Management* 6:421–430.

Sochava, V.B. 1970. The training of geographers for work in applied geography. *Soviet Geography* 11:730–736.

Steers, J.A. ed. 1971. *Applied Coastal Geomorphology.* The MIT Press: Cambridge, MA.

Strahler, A.N. 1952. Dynamic basis of geomorphology. *Bulletin of the Geological Society of America* 63:923–938.

Surell, A. 1841. *Étude sur les Torrents des Hautes-Alpes,* Paris.

Thomas, W. ed. 1956. *Man's Role in Changing the Face of the Earth.* University of Chicago Press: Chicago.

Tinkler, K.J. 1985. *A Short History of Geomorphology.* Barnes and Noble Books: Totowa, NJ.

Utgard, R.O., McKenzie, G.D. and Foley, D. eds. 1978. *Geology in the Urban Environment.* Burgess Publishers: Minneapolis.

Verstappen, H.T. 1983. *Applied Geomorphology: Geomorphological Surveys for Environmental Development.* Elsevier Scientific Publishing: Amsterdam.

Walker, H.J. and Mossa, J. 1986. Human modification of the shoreline of Japan. *Physical Geography* 7:116–139.

Walling, D.E. and Gregory, K.J. 1970. The measurement of the effects of building construction on drainage basin dynamics. *Journal of Hydrology* 11:129–144.

White, G.F. 1966. Deserts as producing regions today. In E. Hills ed., pp. 421–438. *Arid Lands: A Geographical Appraisal.* Methuen and Company: London.

Williams, A.T. and Sothern, E.J. 1986. Recreational pressure on the Glamorgan Heritage Coast, South Wales, United Kingdom. *Shore and Beach* 54:30–37.

Woeikof, A.I. 1901. De l'influence de l'homme sur la terre. *Annales de Geographie* 10:97–114, 193–215.

Zakrzewska, B. 1967. Trends and methods in landform analysis. *Annals of the Association of American Geographers* 57:128–165.

Douglas J. Sherman
Department of Geography
University of Southern California
Los Angeles, CA 90089–0255
U.S.A.

Appraisals from Human Geographers

9. Cultural Geography, Its Idiosyncrasies and Possibilities

More than other subsets of the discipline, American cultural geography has resisted the bandwagons that have swept geography since World War II. Small wonder, then, that its applied dimension has been feeble compared to most other components of the intellectual sprawl that is geography today. The nature of cultural geography as well as the mindsets of its practitioners account for a certain determination and resiliency in the face of strong pressures to follow the herd. This remark is not to suggest, however, that the cultural optic has had no new developments or applications or that all of its researchers fit the same mold. Cultural geography now has a near constellation of thematic concerns, and it has become more difficult to characterize this subfield than it was thirty years ago.

The aim of this essay is to describe some of the rather particularistic assumptions that have characterized cultural geography and its core group of practitioners over the past half century. This background is basic to understanding why attempts to apply this kind of research have been modest, although it tries also to show that cultural geographers deal with a set of topical concerns that have application to national and international problem solving.

Carl Sauer's Influence

The culture concept forms the glue that holds cultural geography together in terms of its content and justification. However, some kinds of research have received much more attention than others, so that cultural geography corresponds more to specific spheres of inquiry than to putative boundaries. The foundation of this endeavor is the work by the Berkeley group that emerged out of a creative contact with anthropology and, to a lesser degree, biology. One remarkable scholar, Carl Ortwin Sauer (1889–1975) stands as the intellectual maestro and inspiration to those who claim affinity to cultural geography (Kenzer 1987). It was Sauer who conceptualized the *Fragestellungen* that led to its formulation as a special subject matter. Without Sauer's scholarly visions, intellectual power, and academic courage to march to a different drummer, cultural geography would very likely not have an identity today in this professional enterprise. Sauer cut himself off from the standing preoccupations of the 1920s and 1930s with their facile determinism, uninspiring commodity studies, and barren regional inventories that were the stock in trade in Midwestern departments.[1] Since Sauer's death, the high level of interest in his writings, life, and philosophy is a measure of his seminal role in the discipline and recognition that his published work continues to be well received. Sauer's skillful integration of environment, biological phenomena, mind, and

M. S. Kenzer (ed.), Applied Geography: Issues, Questions, and Concerns, 135–150.
© 1989 Kluwer Academic Publishers.

cultural tradition into a time-space *Gestalt* was discovered by quite a few people from the late 1960s onward who had no formal connection with geography.[2]

Critical to Sauer was the search for origins or, at least, probing the past, which suggested explanations.[3] In some ways, Sauerian geography merges with the Braudelian sense of history that began to win worldwide recognition in Sauer's final years. Like classic cultural geography, it moves with ease across all frontiers and is open to all techniques (Stoianovich 1976). It extols the *'longue durée'* of geographic time rather than the ephemeral event. It emphasizes the anonymous folk who intervened into the natural order to shape spaces and create cultures each distinct from the other. Indeed, much cultural geography could probably be considered 'history' except that professional historians rarely integrate habitat, culture, and economy. For them, 'geography' is usually an inanimate backdrop that is dismissed after the first paragraph. On another front, cultural geography sometimes overlaps with cultural anthropology, although the work of geographers usually shows evidence of a broader and deeper knowledge of the physical environment.

The actual organization of cultural geography was left to Sauer's intellectual descendants to codify and elaborate. Two of his students of the 1950s, P. Wagner and M. Mikesell (1962), compartmentalized the subfield into five components: culture area, culture history, cultural diffusion, cultural ecology, and cultural landscape. The latter two have subsequently received enough attention to emerge as incipient bodies of knowledge. Geographical cultural ecology touches base with the work of archaeologists and ethnographers much more than with other geographers; indeed, it has become one of the true disciplinary overlaps in the social sciences. It has especially favored preliterate societies in tropical settings and while the time dimension is critical, adaptation seems to be more of a key idea than cultural evolution. The cultural landscape has been a prime subject of reflection, amenable as it is to the continua of rural to urban settlement and preindustrial to postindustrial societies. Physical forces and human actions over time unite to create the observable reality of places. This focus on the visible and concrete became the American geographer's definition of landscape, and here Sauer's influence was decisive. However, Sauer himself seems to have derived his landscape concept from Otto Schlüter (1972–1959), whose research agenda included pioneering work on the *Landschaft* as an assemblage of features apparent to the naked eye (Schick 1982). Cultural landscape as an approach has blossomed since Sauer, for it is a compelling notion to geographers who believe in the primary value of observation. It also has an integrative power that dissolves the false dichotomy between physical and human phenomena. Landscape studies also elucidate cultural lag of features that, while bypassed in time, may in fact represent a set of useful traditions that can be called forth when and if they may be needed (Miller 1971).

Cultural geography has undergone several permutations since the Sauerian era and these have attenuated the emphases of the early Berkeley group. Cultural geographers today no longer have an assumed historical dimension to their work. Moreover, fieldwork is not a foregone conclusion as a research procedure.[4] Another shift is that people now get as much attention as artifacts.[5] The assertion shared by both Vidal de la Blache and Sauer that geography is a science of places and not of people would probably find less general agreement than it did at one time. Furthermore, the amount of research now conducted on modern societies is as much as that carried out on the folkways that intrigued Mr. Sauer. A major new perspective with cultural assumptions, environmental perception, arose in the 1960s to give priority to understanding the actors as decision makers having geographical outcomes.

Cultural geographers have not been eager to philosophize their epistemological stances. What has counted to them is doing geography rather than palavering about its content. The influence of Carl Sauer is apparent here too, for to him the key to good work was to follow one's curiosity unencumbered by 'correct' methodologies. The state of knowledge in a particular subject is always prone to modification and revision to correct facts and develop new hypotheses. Assertions of cause and effect are always tentative, indicative of an idealist position toward knowledge where the aim is to develop a coherent and rational configuration of the world that incorporates both the objective and the subjective. In phenomenology, the subjective is elevated to a higher form of knowing. Sauer (1925, p. 20) used the word phenomenology in at least one of his publications, and his often quoted epigram that natural resources are cultural appraisals reflects his belief of the role of human subjectivity in geography. But Sauer would also have undoubtedly agreed that those same natural resources do have meaning outside the narrow realm of experience. The phenomenological work of cultural geographer Yi-Fu Tuan (e.g., 1984) has promoted a humanistic dimension that has enriched the whole discipline by its cerebral reflections. But Tuan's main inspiration comes from literary allusion, not fieldwork, and in several ways his essays, however beautifully constructed, have lost touch with the earth and people as members of groups.

Any assemblage as resolutely nonconformist as cultural geographers are hard to generalize fairly. But many of them are charmed by diversity and therefore less likely to conceptualize the world in terms of its regularities. The myriad components that make our planet such a complex place become a source of everlasting interest. Cultural geographers trust their senses to make valid observations about patterns and processes. Some researchers (e.g., Simoons 1961; Donkin 1977) have used bibliographic approaches to reveal macro-scale patterns to good advantage by collating large numbers of field observations recorded in the literature. A notable disposition of cultural geographers is their openness to a vast range of data regardless of the discipline that generates it. Also, unlike economic and urban geographers, most cultural geographers did not change their research agendas to fit the Kuhnian model by which one scientific paradigm supersedes another. Classic approaches never disappeared and prior foundations continued to be built upon in any way appropriate to the problem at hand, even with the realization that several important funding agencies counted them out unless they were willing to accept positivistic assumptions and reorder their methodology accordingly.[6] In retrospect, this conservatism has helped to maintain touchstones to the disciplinary past that the enthusiasms of the moment always threaten to discard. Perhaps at the core of this refractory behavior is the cultural geographer's inclination not to place too much store in the mirage of absolute truth, which has become the grand objective of 'pure science.' Rather, sights are set more realistically on advancing knowledge by critically interpreting and reinterpreting both old and new evidence. From this sifting, emendation, and refinement, a more accurate or insightful account emerges.

Aside from its focus on the concrete, the cultural geography tradition has several other hallmarks. The privileged position of fieldwork has been stronger than in any other branch of human geography. Part of this emphasis can be attributed to the mystique of direct contact with the objects of study, but also to the lack of statistical sets for most kinds of cultural data. The manipulations of aggregate data and armchair theorizing that have become the basis for spates of doctoral dissertations in other branches of the field are not usually possible in cultural geography. With its fieldwork tradition, cultural geography can help to revitalize the discipline to pay much more attention to observation. Data bases generated by others can be used in some instances, but dependence on them for anything but the macro-scale impoverishes the field.[7] Too much of our world will be overlooked unless geographers get out and record it first hand. Fieldwork also helps check the temptation to elevate research technique to intellectual content.

Cultural geography also has had less *a priori* bias than most other subfields in the discipline. Ideological entanglements tend to be eschewed, part of a larger rejection of deterministic thinking that makes both environmentalism and Marxism suspect modes in knowing the world. Instead, causation is understood in terms of chance, the result of a fortuitous intersection of an independent series of phenomena. Vidal de la Blache (1979) expressed it succinctly when he wrote early in this century, '*tout ce qui touche à l'homme est frappé de contingence.*' The absence of order that explains the character of places thus becomes an objective reality of the world.

Cultural geography has relevance to a number of topical themes that deal with issues that call for resolution. In some cases, geographers have contributed to these issues; in other instances, the potential exists for doing so because of the nature of their interests, expertise, and experience. A number of cultural geographers have done good work for public agencies or consulting firms that has never been published or widely distributed. What follows in this paper are vignettes of several themes that focus on issues of territory, development, environment, quality of life, and diversity. The cursory treatment of each necessarily leaves many unanswered questions; the point here is to suggest potential avenues for policy-oriented research in cultural geography. It is the possibilities for use in formulating official decisions that the way opens for application of cultural geography.

Culture and Territory

Cultural geographers are the most likely candidates to sort out creatively the culturally derived imperatives of space. Sauer (1941, p. 23), who held political geography in low esteem because of its misapplication in Europe, nevertheless considered competition between cultures for territory to be an appropriate theme for geographical research. Territorial integrity, founded on a sense of cultural distinctiveness from surrounding peoples, is a topic of importance in the organization of every nationally defined space whether or not that nation corresponds to the boundaries of the state.

It is in Québec that geographers have been elaborating the meaning of culture and territory in ways that have challenged people to ponder its application to its geopolitical future. The sovereignty issue that came to the fore in the 1970s triggered much of this thinking. In the 1980s the separatist tide ebbed sharply and has been replaced by a more pragmatic public and private attitude that seeks to take the best from its ties with the Canadian Confederation and Wall Street financiers. Through the vagaries of governmental change, the cultural particularism of Québec persists. The remarkable achievement is the continuity from a seventeenth-century locus of French settlement that failed to dissolve through military conquest by the British in 1759 or the massive Anglo-Saxon cultural influence that has bombarded Québec on three of its sides. What this identity consists of in the late twentieth century, how it should be best described, and how its future can be assured and enhanced, have been subjects of many reflections.

One geographic reality of Québec is its 'immensité et marginalité' (Bélanger 1983, p. 199). Most of this huge chunk of territory is not practical for permanent settlement, so that its control has not been much more than a cartographic exercise. With the progressive abandonment of fringe farming after 1950 and of many mining towns as well, the ecumene has contracted rather than expanded. Filling the boreal void with people is not going to take place in the foreseeable future. But even in the old settled core of the Upper St. Lawrence Valley, occupation of space betrays a certain fragility. The *rang* system, which deployed rural settlement into a Strassendorf type on a cookie cutter model, did its part in thwarting the emergence of a real attachment to the land. Most construction of the present century manifests borrowed architectural styles.[8]

A second reality of geographic import is that the fundamental fabric of Québec society is basically local rather than regional. Although communities vary a good deal in their economic well-being, cohesiveness, and viability, kin and neighborhood ties have been primordial. The provincial level has also been critical because it provided most services and defended the culture. It is the spatial mid-scale between parish and province that is weakly defined. Québécois geographers have started to address the need for a sound regional framework. An interdisciplinary research project from Université Laval, funded by the Québec Ministry of Cultural Affairs, represents a step in this direction (Bureau 1977). It focused on the Charlevoix area north of Québec City and attempted to define the region by aggregating local perception. Nearly 500 different interviews were taped, transcribed, and classified to create perceptual profiles of the character of the Charlevoix landscape, its evolution over time, its presumed individuality, the relationship among groups of people who live there, and how they express a sense of place. A solid foundation for the regionalization of Québec can emerge from research of this sort. Successful planning presumes the use of geographic units considered to be valid entities by the inhabitants themselves. The planning regions used in Québec since the 1970s leave much to be desired as anything but statistical subdivisions. Valid regional categories can have an economic impact on industrial location and tourism. Weakness of the regional image may help to explain why Québécois, avid travelers though they are to other countries, do not visit much of their own province.

Another geographic factor with implications for the Québec future is the lack of a deeply rooted territorial ideology. In the *coureur de bois* tradition, Québécois have shown a good deal of mobility, reflecting a fluid notion of space. Their history had not conditioned them to stay put within their frontiers, a notable contrast with the French of France. When times got tough or opportunities presented themselves elsewhere, the *'départ'* was a safety valve for many individuals. Morrissonneau (1983, p. 16) has called Québécois a *'peuple de passage non de l'enracinement'*; a tightly delimited political entity based on Francophone nationalism could be a straitjacket for their own citizens, however much they may be committed emotionally to their culture.

In the Third World, national borders, often quite recent, do not usually correspond to a homogenous culture. In Africa, a multitude of ethnic groups complicates national life in virtually every country. Since spatial dominance enhances power, central authorities have regarded their European-devised boundaries as sacrosanct. Movements of cultural groups for autonomy or even independence that are likely to emerge in many countries are mercilessly squelched, as occurred in Nigeria during the Biafra struggle. Governments treat their oppression of minority peoples as internal matters and do not wish international publicity. It is left to outside scholars to dig up the facts and unveil the injustices of an imperfect world. Cultural geographers can do their part by speaking out and writing on these issues of cultural survival that span the ideological spectrum. In this way, geography can contribute to informing international opinion, which is often the most effective way of bringing pressure on central authorities to reconsider practices and policies designed to enhance their quest for power and resources.

In Asia, two troubling cases of state expansion are going on under the euphemism of 'national building' (Nietschmann 1986). In Bangladesh, lowland Bengali people have moved with official sanction on to lands traditionally occupied by ethnically different hill tribes. Domestic military assistance has enabled Bengali squatters to dispossess the rightful owners in certain areas. In Indonesia, the territorial agenda is also based on population pressures, but with stronger nationalistic motives than in Bangladesh. There, independence was the culmination of a hard struggle against colonial oppression; hegemony over the entire Netherlands East Indies was seen as the just reward. In the 1970s, this imperialistic design was extended to the invasion of East Timor, a Portuguese colony for more than 400 years. The Timorese conflict resulted in a holocaust of lives and annexation to Indonesia, but received little media attention in the world press, perhaps a factor in the shameful acquiescence of the U.S. government's wish to maintain its working relationship with Jakarta. In the 1980s, Indonesia has moved to incorporate culturally West New Guinea, a territory they had earlier taken from the Netherlands in 1969 and renamed Irian Jaya. The Indonesian government has set out to obliterate Papuan identity in two ways. Javanese, Maduran, and Balinese settlers have been brought in under the national transmigration program, and Papuans are forcibly relocated and indoctrinated into Islam and Javanese culture and language to turn them into 'Irianese.'

Throughout the Western Hemisphere, indigenous peoples have lost some or all of their traditional territory in virtually every country, but it is in those with recent histories of Indian abuse that are of contemporary concern, among them Brazil, Peru, and Paraguay. It is often at the international level where pressure can be exerted effectively to stop the most egregious tyrannies imposed in the name of national destiny. The most controversial Indian rights issue in Latin America in the 1980s has been the fate of the Miskito in Nicaragua. Their treatment, embroiled in the larger conflict between the Sandanista government and the rebels against that regime, transcends political ideologies. Abuses of minority cultures span the continuum from capitalist to communist economies in both developed and under-developed countries. In Panama, a country not ordinarily associated with protection of human rights, a laudible attempt has been made to keep outsiders at bay in Indian-occupied jungles. Two 'cultural parks' have been created, one the Darien World Heritage Site established in 1981, the other the Comarca Emberá-Druá set up in 1983. The intent to maintain the tropical rainforest in those places as the subsistence foundation of their livelihoods is positive; however, the ecological basis is shaky because clustered settlement and growing population have put pressure on local game resources (Herlihy 1986).

Sometimes the territory issue is complicated by migration across political borders. The Garifuna, more commonly known as Black Caribs, whose numbers now surpass 80,000, are found on Dominica, St. Vincent, and the Central American countries of Honduras, Guatemala, Belize, and Nicaragua to which they were exiled since 1797. Their identity rests on a mixed Carib Indian and African race and culture; men fish, women cultivate with an emphasis on manioc cultivation. Most settlements are on or near the beach; housetypes of rectangular form and palm-thatch roof dominate. Davidson's (1984) exhaustive analysis of their complex ethnogenesis establishes the foundations in space and time and forms a document that they can use to confront the governments of countries in which they live as to their legitimacy.

Third World Development and Cultural Geography

Economic development in what we cavalierly call the Third World has been a topic of interest to many cultural geographers with an international perspective. The notion of development with outside funds is still widely interpreted as the way to turn the poor countries of the world around and on the path toward eventual prosperity. Indeed, American perception of other lands rather commonly focuses on their development status, rather than their cultural or environmental character, as the essence of their reality.

The practice of development – a growth industry for the expatriately inclined who are its most ardent supporters – deserves attention from cultural geographers who are willing to look hard and clear at the empirical assumptions on which it is based. Ethnocentric in their design, development schemes based on foreign aid frequently fail. Others that are proclaimed successful usually have negative side-effects – e.g., the expansion of noisome shantytowns on urban fringes as the result of a dam project that dispossesses peasants of their land. Implantation of neotechnic agriculture with tractors, herbicides, and monoculture may actually cause food shortages rather than alleviate them. And development projects have even contributed to disaster. The foreign aid teams that sank tube wells in the Sahel of Africa only dimly perceived that an increased water supply would trigger rapid growth in livestock populations. Excessive numbers of animals devoured the range plants, which resulted in much of the desertification and consequent famine that grabbed world headlines in the 1970s. The cultural ecology of Sahelian herders was essentially overlooked. Droughts have always been part of life in this region, and when they occur herds contract. The Sahel fiasco and others like it suggest that development projects frequently benefit by including native systems of resource management in their plans. The perceptual aspects of these systems may be as important as the 'real' in predicting success or failure of certain projects. Knight (1971) has urged the systematic study of local perceptions of land use and environmental categories – some at the level of the individual, others collective – to serve as a basis for agricultural extension projects and rural development.

Cultural geographers with a holistic viewpoint and no ideological axes to grind can bring a much needed perspective to development projects, especially if they have a complex understanding of the region or locale. Most development 'experts,' local or foreign, tend to work out their ideas in isolation from those they are trying to help and in areas they know little about. The Third World's cultural landscape is littered with white elephants: plans that looked sound on paper, but which failed to take into account local or national conditions. Most foreign economists or agronomists are amateurs in dealing with peasants on their own terms, and most possess a mental block in understanding and appreciating rationales that do not favor economic maximization.

Peasant communities around the world have lost control over their food supplies. One example is in Bolivia, where the government and the U.S.A.I.D. are largely responsible for the parlous state of Bolivian highland agriculture. American aid programs have distributed large quantities of flour for many years. The Bolivian government, for its part, has subsidized the cost of bread to placate city folk. Both policies have discouraged domestic production of wheat, and now less than 20 percent of the Bolivian demand is met by domestic production. By importing flour rather than the whole grain, the local grinding technologies based on water mills have virtually disappeared. Cases like this explain why so many peasants have quit the land and moved to the cities.

Geographers who study single communities have contributed insights to understanding the intricacies of resource use that relate to development. For example, Nietschmann (1979), who studied the village of Tasbapauni, Nicaragua from 1968 to 1976, described how the Miskito Indians there have shown remarkable flexibility in adapting their local economy to changing circumstances as the price of their survival as a group. Grossman (1981) has shown how illusory the benefits were in entering a cash economy based on cattle raising and coffee production in an area of Papua New Guinea with surplus land and labor. In both cases, the micro-scale of analysis allowed for a fuller grasp of the conditions that made the two researchers highly sensitive to the outside factors affecting these village systems. This kind of knowledge is of inestimable value in predicting the success or failure of development projects, and more cultural geographers should get involved at the community level.

Environmental Degradation

The cultural geographer's view of the pervasiveness of human agency has contributed to issues of environmental deterioration. Their concern predates the 'environmental movement' of the late 1960s by at least three decades, and can be attributed to Sauer's rediscovery of the insights of George Perkins Marsh (1801–1882) well before most scholars of his time. Like Marsh, Sauer was concerned with the reasons for such destruction, but unlike Marsh, Sauer was critical of the technocratic attitudes and practices of Western civilization. This concern was cogently elaborated later by W. Clarke (1977) who worked among a subgroup of the Maring people in New Guinea. The paleo-technology that has characterized Papuans for millenia has, above all, a permanence that is impossible with neotechnic introductions. Donald Innis (1983), another Berkeley graduate strongly influenced by the Sauerian land ethos, has written about the superiority of polyculture over the practice of growing but one crop in a field. He has promoted the idea of intercropping as a sound agricultural practice that fosters environmental and economic stability. Whole agro-ecosystems have undergone reevaluation in this regard. Geography textbooks published before the 1960s often described swidden farming as destructive of forests and wasteful of land. Evidence mounts that this land-extensive mode of farming has much to recommend if fallow cycles are not shortened. Among the Bora Indians of eastern Peru, Denevan (1985) and his team found that over 100 different plant species continued to be used after the land had gone to fallow. Field research can invalidate past clichés.

Cultural geographers have documented how people in specific places have changed their habitat through time. Once the time-space dimensions of the processes of destruction are understood, it seems much easier to conceptualize a plan of action. Mikesell (1968) showed how cutting and grazing over many centuries in Lebanon virtually denuded mountains of their once magnificent forests. His frequently cited reconstruction imparts a meaningful perspective that highlights the urgency of protecting the remaining clumps of cedar trees. In another cultural-geographical study, Doughty and Greer (1976) place Chinese wildlife conservation in the context of how that culture has used the native fauna through time.

Human impact on the resource base can be accelerated by the institutionalized peculiarities of land tenure. On the tiny island of Rodrigues in the Indian Ocean east of Mauritius, common ownership of most of the land by a remote governmental administration has promoted the wholesale destruction of soil, vegetation, fauna, and water (Gade 1985). Rodriguans never thought of themselves as custodians of the land they merely occupy, so abuse has been rampant. Rehabilitation of eroded farmland by the British colonial service in the 1950s and 1960s even had to pay peasants to build terraces on their own farm plots – but plots that were leased, not held in fee simple. With population rising above 3 percent a year and crop yields in constant decline, Rodrigues has turned into a neo-Malthusian microcosm of the tragedy of the commons. Vast possibilities exist for cultural geographers wishing to put their special perspective to work on these kinds of topics. Chances are good that their work would form an important exhibit in government dossiers periodically drawn up to address problems demanding attention.

Cultural Geography and the Quality of Life

Cultural geography, even when it is not applied in intent, can be useful in formulating decisions that enhance the livability of our surroundings. In spite of a heavily urbanized demographic profile, Americans are still far from successfully humanizing city life. Larry Ford (1984), one of very few geographers who look at cities from a combined cultural-economic perspective, makes a plea for more study of our urban architectural landscapes to preserve historical buildings. The classifications resulting from these surveys can be useful in planning for change in a city's housing stock, monitoring of redlining practices, construction code enforcement, and rezoning. The major nemesis to the well-being of an urban fabric is the automobile. What we have gained in freedom of mobility we have lost in human scale and shared community, and cultural geographers have had important things to say about the devastating impact of the car on cities. Horvath (1974) showed the large extent to which land formerly designed for people is monopolized for vehicular use. Urquhart's (1981) sage remarks about the wasteland of an automobile strip in Eugene, Oregon can help city officials everywhere in the U.S. think about the need to create a humane environment for people who want alternative modes of locomotion.

The quality of our surroundings, preserving the good and eradicating the bad, is a general societal concern to which cultural geographers can and have addressed themselves. One segment of this large issue is historic preservation, which can be dealt with nicely in the cultural landscape paradigm. Inventory is an essential early task in this effort; by accounting for the location and distribution of cultural features, geography can help to formulate an understanding of the significance of their presence in urban and rural settings. That buildings and other artifacts should be preserved for future generations is a fairly new concept in the U.S., triggered partly by the 1976 Bicentennial and by substantial public funding or tax advantages. Europeans, of course, have long taken for granted that features and even whole landscapes of historic, ethnographic, or folkloric value merit preservation. In a country like Italy, these *beni culturali* as they are called are so common that the task is a difficult one given the ubiquity of lovely old structures and the anarchy of construction since World War II (Ruocco 1979). In the U.S. most people are not yet blasé about the preservation idea and are perceptive to its contribution in creating a richer texture of urban life, if not always a beautiful city. The urban renewal mentality that bulldozed vast sections of American cities has few defenders anymore, and many urban places have become a good deal more livable than they were in the 1960s. Hope reigns even for Cleveland!

Enhancement of Diversity

By their studies of the American ethnic mosaic, cultural geographers can generate knowledge useful in drafting agendas that reinforce national diversity.[9] Cartographic presentation can be of substantial value in showing spatial intensities and borders. Whether cited or not, Nostrand's (1970) map of the Hispanic-American population has probably influenced minority policy formation. For Louisiana, the Protestant north and the Catholic south has always been the most important social division in the state. Two recent maps of the Cajun region (Louder and Waddell 1983; Estaville 1986) can have use in refining cultural boundaries and thus have significance for education and liquor boards.

Another group, now much more spread out than the Cajuns, is the Old Order Amish who maintain quaint ways, at odds within the larger American society (Crowley 1978). Their spatial coordinates have expanded greatly from the South-western Pennsylvania core because of their high birth rate and need for farming land. Landing (1972) found that the interaction of people between Amish settlements does not conform to simple distance. Part of their mobility can be attributed to their 'paradoxical cultural view,' which seeks to take advantage of technology on their own terms. They eschew ownership of cars, but are ready to ride in those owned by non-Amish. Amish do not plow with tractors themselves, but they will hire non-Amish to do the mechanical work. Electricity is *verboten*, but battery-operated flashlights are quite acceptable. Kitchen appliances can be borrowed but not owned. Rather than wash clothes by hand, they bring their horse-drawn buggies to town and use commercial laundermats (Rechlin 1976).

A wide range of cultural data has interpretive value for understanding American diversity. Take rice: American consumption shows strong regionalization (Shortridge and Shortridge 1985). Rice processors and wholesalers concerned with marketing their grain can use this information in their marketing strategies. Diversity takes another form in the study of class differences. Pockets of suburban wealth in the Northeast manifest subtilties of decor that reflect old élitist values drawn from preindustrial Great Britain. Anglophilia is part of this, but so is it a marker that sets these people off from 'new money,' possibly of first or second generation immigrant stock (Duncan and Duncan 1984).

The Future of American Cultural Geography

Partly because the applied dimension is not well developed, cultural geography has colleges and universities as its principal arena of remuneration. For this reason, the subfield has been on a steady downturn for almost two decades.[10] As the quest for 'true science' gained ascendancy in geography, many departments devised job descriptions to hire model builders and theoreticians. Especially in the 1980s, the rage for technical know-how further diminished new appointments in cultural and regional geography. This latter trend has still not run its course in 1987, and corresponds to the larger vocational emphasis that has pervaded American higher education in general. Knowledge not directly applicable to landing a good job is bypassed for subject matter that enhances employability.

Regional geography has eroded even more over the past twenty years than cultural geography, although the two overlap substantially since many cultural geographers use regions as their primary frame of reference. It is regional geography that can be expected to stage a vigorous comeback, and that reawakening should also help cultural geography. A regional perspective can merge the physical and human dimensions, fuse the objective and subjective, and deal felicitously with both pattern and process. It meets an expectation on the part of the general public and the schools who expect geography to be a reservoir of knowledge about the world. With Japan as the successful counterpoint, the realization heightens that a competitive edge in the world marketplace comes from an outward directed society that has a sense of the geographical realities of other countries. Ignorance of potential customers, their economies, politics, resources, and cultures can hobble market penetration in subtle ways. It is curious that today the remote corners of the globe know American popular music, inventions, and generosity, yet that strong influence is not matched by a population with a global sense of human diversity. Only a small minority possess a reasonably accurate mental map to make the vital geographical connections in world affairs. The Vietnam fiasco showed that even our national leaders can be geographical ignoramuses. The Soviet and American contrast forms a striking paradox: although a closed society, the Soviet people show a good deal more interest and general knowledge of the world beyond their borders even though few are given the opportunity to travel abroad. In the U.S., with its open society pegged to individual selffulfillment, most high school and many college students scarcely know where Nicaragua is. As a people, Americans have not opened their minds to other countries or cultures.[11]

Few would deny that knowledge about places is part of our field, however much we have put this aspect of our intellectual patrimony out to pasture over the past two decades. To revive it to its legitimate place in the discipline will take time and determination. Many American geographers under age forty have no in-depth knowledge about the outside world, and the geography lists in *Dissertation Abstracts* suggest that only a handful of the newly-minted American Ph.D.s are capable of filling the breach left by multiple retirements of personnel in regional and cultural geography. Less than a dozen Ph.D.-granting geography departments – those maintaining a strong cultural/regional interest through the past decade – are positioned to train geographers effectively to meet this renewed demand (AAG 1986).

Conclusion

How human agency molds the natural world to create cultural landscapes, how cultures adapt or misadapt to specific environments, and why cultural traits and the people who carry them adapt to specific places and environments are all weighty themes touching the lives of everyone in some way. Yet, the efforts to use cultural geography to solve specific societal problems that relate to these themes have been far overshadowed by the uncontaminated search for 'pure' knowledge. Rather than lament this state of affairs, I do not believe that cultural geographers need be any more applied than they have been. Research that focused primarily on perceived societal needs, especially those on contract funding, would result in a skewed corpus of data about the U.S. (if not the world) that was highly selective and of ephemeral value. If the investigative work of cultural geographers can contribute to the design and implementation of public policy, so much the better. Likewise, the discipline as a whole benefits when cultural geographers, with their formidable breadth of viewpoint, become members or even managers of inderdisciplinary development projects, state commissions, or local advisory boards. But, the choice to set one's research priorities should be paramount, something most scholars in the U.S. have taken for granted. However, quite a few professional geographers trained elsewhere have little or no discretion in formulating their research activities. Third World governments often set up clearly defined expectations in academic research in the interest of national development. Soviet and East European geographers have a much narrower set of problems to investigate than their cohorts in North America. Even in Western societies that are liberal democracies (e.g., the Nether-lands), social pressure plays a role in defining an acceptable geography project. While the smorgarsbord of American geography looks positively anarchical to many, a much greater concern comes from an outside imposition of research expectations that has scholars rushing to get in line. Let us hope that the Sauerian virtue of self-directedness is not lost in the forthcoming generations of American cultural geographers.

Notes

1. Two academic geographers with applied agendas (Frazier 1982, p. 10; Vogeler 1982, p. 285) have referred to Carl Sauer's experience with land use classification in Michigan. Neither mentioned, however, that Sauer repudiated this classificatory work as uninspiring, and his move to Berkeley in 1923 was in part a desire to get a fresh perspective.
2. Most influential outside the discipline of Sauer's published work was his masterful *Agricultural Origins and Dispersals*, which synthesized the notion of domestication in a way that suggests a Goethean inspiration. Goethe believed in a special talent (*'geist'*)

on the part of the researcher to probe beneath the surface attributes of nature to elucidate universal themes (Speth 1987).

3. By way of contrast to Sauer's retrospection, the work of Robert Platt (1891–1964), who also received his Ph.D. from the University of Chicago, had a totally functionalist approach. Platt (1943) made micro-studies at the level of the farm or the town in which he wove the human and physical relationships to exemplify the most 'typical' economic activities at those places. But in these descriptions of the moment, any discussion of how the landscapes had evolved to that point was studiously avoided. By not elucidating change, Platt was not well positioned to say much of worth about the future.

4. Of the ninety-nine articles in the eleven issues of the *Journal of Cultural Geography* (from its inception in 1980 to 1985 inclusive), a serious historical component occupied 63 percent, and a field component 62 percent of the articles.

5. A notable early exception was Edward T. Price's (1953) work on mixed-blood groups in the U.S., whose identities were crystallized by isolation and endogamy brought on by racial segregation.

6. Buttimer (1974), in one of the more eloquent and probing reflections produced by a geographer, suggests that positivism reigned supreme in American geography as long as no cracks appeared in the prevailing WASP ethos that has dominated the field. Her formation as a Dominican nun, since abandoned, prepared her to understand the world in other than the mechanistic terms that lead so comfortably into positivist assumptions.

7. An interesting example of creative thinking outside the realm of the official statistics is Hart's (1977) study of land rotation in the southern Appalachians in which crop land is intercalated with land in brush and woods. Agricultural censuses did not deal with this phenomenon and most agricultural economists have not deemed it worthwhile to investigate 'backward' farming practices. What is officially called land abandonment is more properly a long-term fallow.

8. One can, however, see the effects of a sensitivity to cultural roots in the large numbers of Normandy-type houses with gable windows and steep-pitched roofs (well adapted to heavy snowfall) that have been constructed, especially since 1965. Another indication of change in Québec is the vigorous attention given to toponomy as a way of reasserting the cultural ties with place.

9. A problematic side to the ethnicity question is its use by the U.S. federal government ostensibly to help disadvantaged Americans achieve more opportunity. But it has put people into Manichean categories, both in terms of race and national origin. The descendants of Spanish stone cutters from Santander who came to Barre, Vermont a century ago have nothing in common with the Mexican immigrants in East Los Angeles except that their last name may be Gutierrez.

10. Cultural geography as a declared proficiency among members of the Association of American Geographers fell 25 percent between 1978 and 1986. But, it still remained the third largest topical specialization in 1986 (AAG 1987, p. 35).

11. Most geographers recall the disparaging remarks made about the *National Geographic Magazine* during their graduate school days: packaged for a mass audience, no depth, desultory prose. Yet geographers have a marvellous ally in the National Geographic Society and every one of their publications in educating the public about the world, which, in many, often subtle, ways helps the discipline. Their magazine is a fine popular publication; the quality of the graphics is unexcelled and the accuracy of the information is higher than that found in the *Annals of the A.A.G.*, for each published fact is checked by staff members against two or more sources.

References

AAG 1986. *Guide to Departments of Geography in the United States and Canada, 1986–1987*. Washington, D.C.: Association of American Geographers.

AAG 1987. 1986 AAG topical and areal proficiencies. *AAG Newsletter* 22(1):35.

Bélanger, M. 1983. Architectures, la culture dans l'espace: un certain contexte. *Questions de Culture* (Montréal) 4:199–208.

Bureau, L. 1977. Des paysages, des idées et des hommes: le projet collectif de Charlevoix. *Cahiers de Géographie de Québec* 21:187–220.

Buttimer, A. 1974. *Values in Geography*. Commission on College Geography Resource Paper No. 24. Washington: Association of American Geographers.

Clarke, W.C. 1977. The structure of permanence: the relevance of self-subsistence communities for world ecosystem management. In *Subsistence and Survival: Rural Ecology in the Pacific*, edited by T. Bayliss-Smith and R. Feachem, pp. 363–384. New York: Academic Press.

Crowley, W.K. 1978. Old Order Amish settlement: diffusion and growth. *Annals, Association of American Geographers* 68:249–264.

Davidson, W.V. 1984. The Garifuna in Central America: ethnohistorical and geographical foundations. In *Black Caribs: A Case Study in Biocultural Adaptation*, edited by M.H. Crawford, pp. 13–35. New York: Plenum Press.

Denevan, W.M. et al. 1985. Indigenous agroforestry in the Peruvian Amazon: Bora Indian management of swidden fallows. In *Man's Impact on Forests and Rivers* (Vol. I of *Change in the Amazon Basin*), edited by J. Hemming, pp. 137–155. Manchester: Manchester University Press.

Donkin, R.A. 1977. *Agricultural Terracing in the Aboriginal New World*. Tucson: University of Arizona Press.

Doughty, R.W. and Greer, C.E. 1976. Wildlife utilization in China. *Environmental Conservation* 3:200–208.

Duncan, J.S. and Duncan, N.G. 1984. A cultural analysis of urban residential landscapes in North America: the case of the Anglophile elite. In *The City in Cultural Context*, edited by J. Agnew, J. Mercer and D. Sopher, pp. 255–276. Boston: Allen and Unwin.

Estaville, Jr., L.E. 1986. Mapping the Louisiana French. *Southeastern Geographer* 26:90–113.

Ford, L.R. 1984. Architecture and geography: toward a mutual concern for space and place. *Yearbook, Association of Pacific Coast Geographers* 46:7–33.

Frazier, J.W. 1982. Applied geography: a new perspective. In *Applied Geography: Selected Perspectives*, edited by J.W. Frazier, pp. 3–22. Englewood Cliffs: Prentice-Hall.

Gade, D.W. 1985. Man and nature on Rodrigues: tragedy of an island common. *Environmental Conservation* 12:207–216.

Grossman, L. 1981. The cultural ecology of economic development. *Annals, Association of American Geographers* 71:220–236.

Hart, J.F. 1977. Land rotation in Appalachia. *Geographical Review* 67:148–166.

Herlihy, P.H. 1986. Indians and rainforest collide – the cultural parks of Darien. *Cultural Survival Quarterly* 10:57–61.

Horvath, R.J. 1974. Machine space. *Geographical Review* 64:167–188.

Innis, D. 1983. Aspects of Jamaican post-industrial agriculture. *Journal of Geography* 82:222–226.

Kenzer, M.S. ed. 1987. *Carl O. Sauer – A Tribute*. Corvallis: Oregon State University Press for the Association of Pacific Coast Geographers.

Knight, C.G. 1971. Ethnogeography and change. *Journal of Geography* 70:47–51.

Landing, J.E. 1972. The Amish, the automobile, and social interaction. *Journal of Geography* 71:52–57.

Mikesell, M.W. 1968. The deforestation of Mount Lebanon. *Geographical Review* 58:1–20.
Miller, Jr., V.P. 1971. Some observations of the science of cultural geography. *Journal of Geography* 70:27–35.
Morrissonneau, C. 1983. Le peuple dit ingouvernable du pays sans bornes: mobilité et identité québécoise. In *Du continent perdu à l'archipel retrouvé*, edited by D.R. Louder and E. Waddell, pp. 11–23. Québec: Presses de l'Université Laval.
Nietschmann, B. 1979. Ecological change, inflation, and migration in the Far West Caribbean. *Geographical Review* 69:1–24.
Nietschmann, B. 1986. Indonesia and Bangladesh: economic development by invasion of indigenous nations. *Cultural Survival Quarterly* 10:2–12.
Nostrand, R. 1970. The Hispanic-American borderland: delimitation of an American culture region. *Annals, Association of American Geographers* 60:638–666.
Platt, R.S. 1943. *Latin American Countrysides and United Regions*. New York: McGraw-Hill.
Price, Jr., E.T. 1953. A geographic analysis of white-Indian-Negro racial mixtures in eastern United States. *Annals, Association of American Geographers* 43:138–155.
Rechlin, A.T.M. 1976. *Spatial Behavior of the Old Order Amish of Nappanee, Indiana*. Michigan Geographical Publications No. 18. Ann Arbor: Department of Geography, University of Michigan.
Ruocco, D. 1979. Beni culturali e geografia. *Studi e Ricerche di Geografia* (Genoa) 2:1–16.
Sauer, C.O. 1925. The morphology of landscape. *University of California Publications in Geography* 2:19–54.
Sauer, C.O. 1952. *Agricultural Origins and Dispersals*. New York: American Geographical Society.
Sauer, C.O. 1941. Foreword to historical geography. *Annals, Association of American Geographers* 31:1–24.
Schick, M. 1982. Otto Schlüter 1872–1952. *Geographers: Biobibliographical Studies* 6:115–122.
Shortridge, J.R. and Shortridge, B.G. 1983. Patterns of American rice consumption 1955 and 1980. *Geographical Review* 73:417–429.
Simoons, F.J. 1961. *Eat Not This Flesh: Food Avoidances in the Old World*. Madison: University of Wisconsin Press.
Speth, W.W. 1987. Historicism: the disciplinary world view of Carl O. Sauer. In *Carl O. Sauer – A Tribute*, edited by M.S. Kenzer, pp. 11–39. Corvallis: Oregon State University Press for the Association of Pacific Coast Geographers.
Stoianovich, T. 1976. *French Historical Method: the Annales Paradigm*. Ithaca: Cornell University Press.
Tuan, Y.F. 1984. *Dominance and Affection: The Making of Pets*. New Haven: Yale University Press.
Urquhart, A.W. 1981. Stripping the urban landscape. *Yearbook, Association of Pacific Coast Geographers* 43:7–22.
Vidal de la Blache, P. 1979. *Tableau de la géographie de la France*. (reprint of 1903 edition) Paris: J. Tallandier.
Vogeler, I. 1982. Applied rural geography: choices and opportunities. In *Applied Geography: Selected Perspectives*, edited by J.W. Frazier, pp. 283–304. Englewood Cliffs: Prentice-Hall.
Wagner, P.L. and Mikesell, M.W., eds. 1962. *Readings in Cultural Geography*. Chicago: University of Chicago Press.

Daniel W. Gade
Department of Geography
University of Vermont
Burlington, VT 05405
U.S.A.

10. Applied Recreation Geography

Recreation geography can be loosely defined as '...the systematic study of recreation patterns and processes in the landscape' (Smith 1983, p. xiii). However, the subdiscipline investigates leisure-related phenomena in general, and recreation geographers have investigated the geographic aspects of such diverse phenomena as international travel and tourism (Boniface and Cooper 1987; Matley 1976), collegiate sports recruiting (Rooney 1980), urban recreation systems (Lovingood and Mitchell 1978), and recreational boating patterns (Heatwole and West 1982).

Accommodating the tremendous diversity of these and other leisure-related areas of interest might eventually provoke a change of the subdiscipline's name – perhaps to 'leisure geography.' Meanwhile, recreation geographers commonly think of their work as falling under the general heading of 'recreation, tourism, and sport' (RTS), which is the name of the Association of American Geographers (AAG) Specialty Group that promotes recreation geography interests. For most practical purposes recreation geography can be conceptualized as the leisure-related or 'RTS-themed' works of geographers.

The geographic literature already offers a number of works describing what recreation geography has been and is now. This area of specialization was recently assessed from the overview perspective by Lisle Mitchell and Richard Smith in 'Recreation Geography: Inventory and Prospect' (1985). Stephen Smith explored its roots and origins in 'Reflections on the Development of Geographic Research in Recreation: Hey Buddy, Can You S'paradigm?' (Smith 1982), and carefully delineated major research themes and needs in *Recreation Geography* (Smith 1983). J. Coppock (1980, 1982) has provided an excellent overview of the leisure-related literature contributed by British geographers. Since tourism studies dominate the recreation geography literature (Mitchell 1981; Cheek 1987), it is not surprising that works of that type have been reviewed and discussed in considerable detail (Mitchell 1984, 1987). Our purpose here is not to review and update these and similar works (e.g., McMurry and Davis 1954; Butler 1982); rather, it is to evaluate recreation geography's role in leisure-related problem solving and to assess the effects of the applied geography movement on the future of this geographic research specialization.

Research Orientations

Applied recreation geography is a rather nebulous concept. It can be operationally defined as RTS-themed research conducted specifically to meet the problem solving or policy- and decision-making needs of particular agencies, firms, or

M. S. Kenzer (ed.), Applied Geography: Issues, Questions, and Concerns, 151–163.
© 1989 *Kluwer Academic Publishers.*

programs. In actuality, the influences of RTS-themed studies on managerial thinking, facilities development, marketing, and other behavior are difficult to assess because they are commonly indirect and nonobvious. Perhaps it is best to think of applied recreation geography in the broad context of geographic expertise used as a tool for solving leisure-related problems.

Although recreation geography is a very heterogeneous field, the problem solving aspects of the subdiscipline's contributions can be described fairly efficiently in terms of primary research orientations. Recreation geography functions at all four levels of scientific inquiry – i.e., it yields descriptive, explanatory, predictive, and normative findings. Since some of the most pragmatic information furnished by recreation geographers answers questions about the what, where, who, when, how, and why of leisure-related behavior, the contributions of descriptive and explanatory works are very important. It is the predictive and normative works that lie solidly within the domain of problem solving, however, since the former yields forecasts and the latter furnishes policy guidance.

With few exceptions, research in recreation geography addresses topics that can be allocated to just two major categories: travel and location. Stephen Smith (1983) has devised a typology of recreation geography research that integrates the travel and location themes at all four levels of inquiry. From this perspective, what applied recreation geography has to offer consists of descriptive, explanatory, predictive, and normative research investigating leisure-related travel, plus descriptive, explanatory, predictive, and normative research focused on the locational aspects of leisure-related behavior.

Smith's detailed examination of recreation geography research themes illustrates clearly that research of this genre can take a great variety of forms. Descriptive research focused on recreation resource and activity locations yields location description and resource inventory or other types of descriptions, such as the images people have of recreation regions, landscapes, and facilities. Geographers have used this approach to study recreational climates (Crowe, McKay, and Baker 1977), recreational land use capability (Coppock, Duffield, and Sewell 1974), the recreational potential of rivers (Chubb and Bauman 1977), park use (Smith and Smale 1980), rural festivals (Janiskee 1980), family campgrounds (Janiskee and Lovingood 1986, 1987), vacation region images (Goodrich 1978), landscape esthetics (Leopold 1969), sports landscapes (Raitz 1987), and other leisure-related phenomena. Many worthwhile research topics of this genre – resort morphology, for example (Stansfield 1971) – languish for want of attention.

Descriptive research on recreational travel investigates 'nodes, roads, and modes' – i.e., recreationist origins and destinations, travel routes, and means of transportation. Roy Wolfe's (1976) classic study of summer weekend recreational travel, for example, entailed the plotting and mathematical projection of traffic volumes on secondary roads in the province of Ontario.

Explanatory research on location investigates the location, development, and/or use of public and private recreational facilities. Thus, for example, Lisle Mitchell and Paul Lovingood (1976) studied urban park density in relation to population characteristics in Columbia, South Carolina, while Richard Hecock (1970) studied beach use patterns and Niels West and Charles Heatwole (1979) studied ethnicity and attitudes as factors influencing urban beach use. Explanatory research on travel seeks to understand how specific 'push' and 'pull' factors combine to yield actual patterns of recreational travel, including route and mode choices. Falling into this category are studies investigating the influences of fuel cost and availability on travel behavior (Cosi and Harvey 1980; Wolfe 1980) and other recreational travel-related works, such as the study of mass transit in relation to beach access in the metropolitan New York area (Heatwole and West 1980).

Predictive research on location and travel is aimed at building models and generating forecasts pertinent to selecting recreation facility sites, projecting the movements of recreationists (Van Doren 1967), and performing services such as comparing recreation management alternatives (Lieber and Fesenmaier 1984), forecasting boating trends (West and Heatwole 1979), and predicting the potential impacts of changes in recreational climates (McBoyle and Wall 1987; McBoyle et al. 1986; Wall et al. 1986a, 1986b). Employing transportation cost, locational interdependence, or generalized market locational theories, recreation geographers have sought to determine which locations have situational characteristics inherently best suited for particular types of recreation facilities, services, or enterprises. The development and refinement of trend analysis, gravity, inertia, systems, intervening opportunity, and other models aimed at predicting recreational travel behavior continues to be a major area of research interest (Smith 1983, pp. 121–151; Fesenmaier and Lieber 1985; Cordell et al. 1985).

Clearly demonstrating what recreation geography is 'good for,' normative studies address the question of which locations are best for specific leisure facilities or enterprises, and which levels and kinds of recreation, tourism, and travel development yield the most desirable (or least undesirable) economic, social, and ecological effects. Site selection techniques based on the analysis of social, physical, economic, and transportation characteristics have been used in feasibility studies pragmatically seeking the best sites for vacation homes (Ragatz 1974), campgrounds, golf courses, winter resorts, marinas, sports franchises, and other leisure-related phenomena. A related body of works focuses on selecting sites for public recreation facilities and services in order to satisfy unmet needs, maximize recreational opportunities, minimize average or maximum travel distances, reduce development costs to acceptable levels, and meet associated objectives. Recreation geographers are also helping to answer questions about how recreation-, tourism-, travel-, and sports-related enterprises, facilities, activities, and environmental resources can be developed and managed to yield the greatest social and economic benefits with the least practicable damage to indigenous cultures, natural systems, and recreational quality (Schoolmaster 1986; Murphy 1985; Schoolmaster and Frazier 1985; Nichols 1980; Stanley and Baden 1977; Stankey 1973; Heatwole and West 1982; Parsons 1973; Shechter and Lucas 1978).

Inventorying Relevant Works

No comprehensive inventory of applied work in recreation geography has been conducted, so it is difficult to say precisely how many of which types of problem-directed research have been completed where, why, and for whom. One complicating factor is that much RTS-themed research exists in the form of feasibility studies conducted under contract for clients who want to protect the confidentiality of the information. Another factor is the difficulty inherent in trying to make delicate distinctions between 'applied' and 'pure' research. A significant number of the published works of an applied nature are easy to identify as such because they are agency reports, monographs, symposium papers, or the like contributed by geographers employed by government agencies or working under contract to them (e.g., Knopf and Lime 1984). It remains that most articles in professional journals do not clearly indicate whether the research involved was conducted for a specific agency or firm. The bulk of the information about applied work in recreation geography is provided in paper or panel discussion sessions at the annual AAG meetings, the annual Applied Geography Conferences, and similar professional meetings, or circulates via informal contacts involving discussions of research contracts, formal or informal consulting arrangements, internships, student team projects, unpublished reports, and other projects and products of the problem solving variety. The Specialty Group's *RTS Newsletter*[1] also mentions RTS-themed works of the applied variety.

Fortunately, some special efforts are now being made to assess the status and prospects of applied work in the subdiscipline. Lisle Mitchell and Richard Smith included a section on applied recreation geography in their inventory and prospect article for the *Professional Geographer* (1985, pp. 11–13). More recently, Richard Smith commented on the status of applied work during a panel discussion session at the 1987 annual meeting of the AAG (Smith 1987b). Lamenting the scarcity of reliable information about applied work in the RTS field, he has used the *RTS Newsletter* (Smith 1987a) and other means to ask recreation geographers to send him information about their applied work in order to compile an inventory for periodic dissemination to the RTS membership and publication in appropriate professional journals.

Niches for Recreation Geography

Recreation geographers generally agree that applied work in their field is too small in volume and growing at too slow a pace. Still, there is no question that works of this genre have filled needs existing at all levels of government and within several promising niches in the private sector. Beginning in the 1960s, and until it was virtually dismantled in the 1980s, the recreation geography research arm of the United States Forest Service – represented most notably by Robert Lucas, David Lime, and George Stankey – made an impressive series of contributions to the scientific management of public recreation resources. Cited extensively in the leisure science literature, their contributions have been prominent especially in the areas of carrying capacity analysis (Stankey 1973), recreation choice modeling (Knopf and Lime 1984), park and wilderness recreation management (Shechter and Lucas 1978; Stankey and Baden 1977; Peterson, Anderson, and Lime 1982), and the recreational use of river corridors (Lime and Field 1981).

Over the past several decades academic geographers working in five different states (Ohio, Illinois, Indiana, Texas, and Oklahoma) have executed user survey, modeling, and special studies projects in direct support of statewide comprehensive outdoor recreation planning (SCORP) efforts. The SCORP projects have been significant particularly because they have yielded numerous user survey, demand modeling, and associated special studies vividly demonstrating that recreation geography research has an important role to play in recreation planning.

Some recreation geographers have also worked on behalf of local and regional governments addressing parks management, facilities siting, public resources access, recreation services delivery, and related problems. The *Flower Mound Parks and Recreation Study* (Schoolmaster and Nelson 1984), for example, was conducted by geographer Andrew Schoolmaster and his colleagues to obtain information vital to the park and recreation planning and development processes in the community of Flower Mound, Texas. Charles Heatwole and Niels West have conducted Sea Grant-funded and related research projects investigating opportunities and problems associated with the management of recreational boating, shorebased fishing, and beach use in the metropolitan New York area (Heatwole and West 1980, 1982).

Sports geographer John Rooney has been asked frequently to serve as a consultant to professional sports associations, sports equipment manufacturers, resort and sports facility developers, and other private sector interests. A recent study yielded detailed information about the asymmetrical distribution of golf courses with respect to markets. After presenting some of his findings at a National Golf Association forum, he was besieged with more than 100 phone calls from firms and organizations desiring consulting services (Rooney 1987).

Other instances of distinguished service on the applied geography front can be cited. For example, Gilbert Grosvenor, President of the National Geographic Society, served as Vice-Chairman of the prestigious President's Commission on Americans Outdoors (PCOAO 1987), and geographer Carlton Van Doren has made so many valuable contributions to the fields of recreation, tourism, and leisure research, including applied works of various types (e.g., Van Doren 1975, 1984), that he has been honored formally by the National Parks and Recreation Association.

Canadian geographers account for a disproportionately large share of North America's recreation geographers – about 40 percent – and a similar share of the RTS-themed studies that are oriented to problem solving. Roy Wolfe, Stephen Smith, Geoffrey Wall, Peter Murphy, Richard Butler, and other Canadian recreation geographers have conducted a considerable amount of research and furnished consulting services on a frequent basis to provincial government agencies, such as the Ontario Ministry of Culture and Recreation and the Art Gallery of Ontario, as well as for regional or district agencies, such as Vancouver Island's Cowichan Valley Regional District (Murphy and Andressen 1985) and national agencies and commissions like Environment Canada (Wall 1985), Parks Canada, Statistics Canada, and the National Task Force on Tourism (Murphy 1987).

It is abundantly clear that the demand for information and insights of the type that recreation geographers can provide is growing at a very rapid pace. Tourism is becoming much more economically, socially, and environmentally significant as the world economy undergoes restructuring from an urban-industrial base to an information base. According to World Tourist Organization estimates, world travel expenditures in 1981 stood at about $919 billion (Murphy 1985, p. 3). Factors such as increasing acceptance of leisure-centered and fitness-wellness lifestyles are also helping to shove leisure to the fore as a subject worthy of serious study. Recreation geographers know their field is being taken more seriously now, and that they will be expected to play a significant role in all sorts of policies and decisions related to leisure problems.

There are some good reasons to believe that recreation geography will forge ahead in the problem solving domain and join the ranks of subdisciplines widely known and respected for the volume and quality of their applied works. However, there are also some factors and trends to suggest that this will not happen easily or quickly, and conceivably might not happen at all. Recreation geography is not a principal driving force for the boom in leisure research, nor are its inputs widely regarded as unique or indispensable.

Limitations

A variety of factors have limited recreation geography's contributions to the applied geography movement. Published works and papers presented at professional meetings reveal that not only is the subfield nebulously defined and topically diverse, but it also lacks a core interest, methodology, and philosophy, is excessively oriented to empirical-descriptive-ideographic studies, and is evolving with no clear goals in mind. As Stephen Smith (1982, p. 19; 1983, pp. 183–190) and Lisle Mitchell (1985) have concluded independently, recreation geography is actually in an immature, 'pre-paradigmic' state because it lacks a 'great question' to establish a sense of purpose and direction, a research agenda for the allocation of the subdiscipline's resources, and a unified set of theories and methods to insure scientific rigor.

Recreation geography has not existed very long as a distinct subdiscipline. In 1964, when Roy Wolfe – the 'godfather of recreation geography' – called for a substantial expansion of leisure-related geographic research (Wolfe 1964), only a handful of geographers in North America were recreation research specialists, and there was no true identity for this subfield except as an appendage of economic geography (McMurry and Davis 1954). In subsequent years, recreation geography has received recognition as a distinct subdiscipline, developed formal organizational structures at the national and international levels – including the International

Geographical Union (IGU) Commission on Leisure and Tourism – and generated a readily distinguishable corpus of literature. Nevertheless, even though recreation-related geographic research is voluminous, has a long history, and is firmly rooted in the area studies, earth science, human-land, and spatial traditions of geography (Smith 1982), the recreation geography research specialty is quite young and still evidences the inefficient growth that is a distinguishing characteristic of systems in their early development stages.

Despite statistics that imply otherwise, the cadre of practicing specialists has remained small. In the mid-1980s, new recreation geographers were being trained in only a few universities, and there were still only about fifty-nine academic geographers (scattered among twenty-seven geography departments in the U.S. and sixteen in Canada) whose research interests and publications reflected a strong and continuing orientation to recreation, travel, tourism, or sport themes (Mitchell and Smith 1985, pp. 10–11). Yet, 118 RTS-themed theses and dissertations were written during the 1970s alone, and at least 162 articles and 288 papers on RTS topics were produced between 1962 and 1982. In addition, by 1987 there were nearly 200 members in the RTS Specialty Group of the AAG, or about the same number as in the Climatology Specialty Group. In recent years very few of the more than three dozen other AAG Specialty Groups have sponsored more special sessions at the annual AAG meetings. Even in the realm of applied geography per se, recreation geography is well represented. For example, thirty RTS-themed papers have been presented at the Applied Geography Conferences. The program of the 10th Annual (1987) Applied Geography Conference included two RTS-themed paper sessions: one on tourism in the Appalachian region, the other on a potpourri of recreation topics (Frazier et al. 1987).

This volume of output and participation is possible not only because some recreation geographers are very prolific writers, but also because recreation geography is a field of study replete with 'dabblers' – i.e., people who do not consider themselves recreation geographers, but contribute a thesis or dissertation on a leisure-related topic, or perhaps one or two RTS-themed articles or papers. Mitchell and Smith (1985, p. 10) reported that 86 percent of the individuals authoring or coauthoring articles and 77 percent of those presenting or coauthoring AAG meeting papers between 1962 and 1982 made only a single contribution of the RTS-themed variety. Such contributions, although welcome, are not a satisfactory substitute for the output of a substantial number of specialists doing the scientific-theoretical-nomothetic research required for steady progress in applied recreation geography.

Unfortunately, recreation geography has thus far made little appreciable impact on theory and methodology in the broad field of leisure-themed research. In his commentary on 'The Tao of Recreational Geography,' Lisle Mitchell pointed out that research in the subfield is dominated by descriptive studies of the unique case and generally much too empirical in nature (Mitchell 1985, p. 149). He concluded that recreation geographers must become considerably more scientific if they are to make rapid progress toward understanding what is manifestly very complicated subject matter. Since scientific-theoretical-nomothetic research is the bedrock of applied geography, recreation geography's participation in the applied geography movement has been hampered in a very fundamental way.

Inadequate or missing data bases are also a perennial source of difficulty. Data bases for national recreation participation are now updated routinely at about five-year intervals in the United States (Cordell et al. 1985). However, not since the halcyon days of the Outdoor Recreation Resources Review Commission studies in the 1960s has good quality leisure-themed data been available readily at the local, state, regional, and national levels. Forced to gather their own data for the most part, recreation geographers have opted often to study unique cases at smaller scales so that the data can be acquired with comparative ease.

Another major reason that applied recreation geography has fallen far short of realizing its potential is that recreation geographers – especially those in the U.S. – have done a generally poor job of disseminating the results of their research, cultivating new markets for their information and expertise, and identifying sources of data and funding for their work (Schoolmaster 1987). Many recreation geographers do not choose to publish their findings in the nongeographic journals routinely read by recreation researchers, planners, and decision makers. Few recreation geographers have presented papers at nongeographic professional meetings – such as the National Parks and Recreation Association or the Travel Research Association – sought election or appointment to influential boards or commissions, or otherwise evidenced enthusiasm for carrying the message of recreation geography's worth beyond the confines of academic geography forums. With a few conspicuous exceptions (e.g., sports geography), so little effort has been directed toward cultivating the private sector as a consumer of recreation geography research that this market is scarcely served at all.

Having relied excessively on public agency sponsorship and funding for its research, recreation geography has been hurt severely by the decline in state and federal support for leisure-themed research (Smith 1987). Changes in the nature of statewide planning required for Land and Water Conservation Fund eligibility, crucial for the conduct of SCORP studies, have resulted in an absolute decline in the volume of applied recreation geography research performed under contract for state governments. Federal administration policies implemented during the 1980s under Ronald Reagan also yielded large cutbacks in the funding and staffing of several federal agencies employing recreation geographers (Smith 1987b).

Conclusions

This is certainly an interesting and exciting time to be a recreation geographer. After a slow start, the subdiscipline has achieved a critical mass and seems destined to enjoy a bright future. There are many reasons for this, but none more important than the abundance of opportunities to extrapolate the results of basic research into planning and policy- and decision-making contexts in the nonacademic sector. These opportunities are of many types, and seem especially numerous in the areas of tourism development, recreation and tourism planning, and sports.

There is no question that the application of recreation geography knowledge and expertise to problem solving contexts outside academia offers potential rewards of considerable worth to the subdiscipline: more jobs for recreation geographers, a stimulus to academic research with implications for problem solving, a more clearly defined sense of purpose or social worth, and greater visibility both within and outside academic circles. Since there is no clear distinction between 'basic' and 'applied' research, nor any appreciable threat to quality scholarship, there is no simmering argument on the issue of whether applied research is good for recreation geography. Rather, the real question is whether recreation geographers will have the resources and the zeal to move into the problem solving domain on a much more widespread and consistent basis.

The future of recreation geography as an applied science will be determined by progress in four areas of major concern. One is the need for a unifying paradigm for recreation geography research. There must be a generally accepted research agenda, and this requires the clear delineation of central questions to guide RTS-themed geographic research. A consensus probably will emerge over the course of time, but the subdiscipline's major organizations – the AAG's Specialty Group on Recreation, Tourism, and Sport, and the IGU's Commission on Leisure and Tourism – could do much to accelerate and systematize the process.

Another major area of need is the utilization of more and better geographic tools – including appropriate concepts, theories, and models – for leisure-related problem solving. Recreation geographers should join other leisure research specialists in urging local, state, and federal government agencies to gather and disseminate more information of the type they want, but they will still be hampered for a long time to come by the sporadic availability and generally poor quality of data related to leisure behavior. Nevertheless, there is no fundamental reason why there should not be progress substantially greater than what now exists on the theory- and model-building fronts. Recreation geographers can pave the way to vastly improved problem solving capabilities by shifting attention away from the unique or ideographic and placing more emphasis on the predictive and normative approaches to research. This shift will necessarily entail the use of more sophisticated descriptive and analytical methods – such as numerical modeling and computer cartographic techniques, geographic information systems, and remote sensing – as well as approaches that are interdisciplinary, multidisciplinary, and synthesizing in nature. It might be argued that this perspective and its associated research techniques should permeate the academic training of most, perhaps all, recreation geographers.

A third important need is for heightened awareness and appreciation of problem solving needs and opportunities outside the traditional bounds of scholarly research. In the United States, and to a lesser degree in Canada, recreation geography evolved in an academic environment that viewed quality scholarship per se as the principal product of geographic endeavor. Accordingly, recreation geographers have been disinclined to stray from basic research except in special circumstances. Now, it will be necessary to accord applied recreation geography considerably greater status, with commensurate recognition and other rewards for its practitioners, if recreation geography is to realize its potential as a problem solving area of expertise. When this is achieved, many more recreation geographers will come to see applied work as a core concern rather than as a potentially lucrative sideline, and problem solving needs in the nonacademic world will come to be seen as the root basis for numerical and qualitative growth in the subdiscipline.

A final important need is for the aggressive marketing of recreation geography expertise in the nonacademic realm (Abler 1987). The world is not going to beat a path to the subdiscipline's door and demand better access to the collective genius. Indeed, it is probably fair to say that most planners and managers in the leisure-related fields have never heard of recreation geography; nor do they care if it exists. If there are going to be more 'niche-spaces' for recreation geographers in the nonacademic world, recreation geographers themselves must recognize the opportunities and effectively compete for them. Since the subdiscipline is small, overwhelmingly academic in its background and orientation, and mostly comprised of individuals who have thus far seen little need to colonize new turf, it is quite possible, perhaps even probable, that applied recreation geography will not blossom dramatically; instead it will grow in the slow but steady way that recreation geography itself has grown over the past several decades.

Note

1. The *RTS Newsletter* is the official newsletter of the AAG Specialty Group on Recreation, Tourism, and Sport. Usually published two or three times per year, it is distributed to members of the Specialty Group at no additional cost.

References

Abler, Ronald F. 1987. What shall we say? To whom shall we speak? *Annals of the Association of American Geographers* 77(4):511–524.

Boniface, Brian G., and Cooper, Christopher P. 1987. *The Geography of Travel and Tourism*. London: Heinemann.

Butler, Richard W. 1982. Geographical research on leisure – reflections and anticipation on Accrington and fire hydrants. In D. Ng and Stephen Smith (eds.), pp. 43–67. *Perspectives on the Nature of Leisure Research*, Waterloo, Canada: The University of Waterloo Press.

Carlson, Alvar V. 1980. A bibliography of geographical research on tourism. *Journal of Cultural Geography* 1:161–184.

Cheek, William, ed. 1987. Annals of tourism research. *RTS Newsletter* (July): 5–6.

Chubb, Michael, and Bauman, E.H. 1977. Assessing the recreation potential of rivers. *Journal of Soil and Water Conservation* 32:97–102.

Coppock, J.T. 1980. The geography of leisure and recreation. In *Geography: Yesterday and Tomorrow*, pp. 263–279. London: Oxford University Press.

Coppock, J.T. 1982. Geographical contributions to the study of leisure. *Leisure Studies* 1:1–27.

Coppock, J.T., Duffield, B., and Sewell, D. 1974. Classification and analysis of recreation resources. In P. Lavery, (ed.), pp. 231–258. *Recreational Geography*, London: David and

Charles.

Cordell, H. Ken, Fesenmaier, Daniel R., Lieber, Stanley R., and Hartmann, Lawrence A. 1985. Advancements in methodology for projecting future recreation participation. In *Proceedings of the 1985 National Outdoor Recreation Trends Symposium II, Volume I*, pp. 89–109. Atlanta, GA: USDI National Park Service, Southeast Regional Office.

Corsi, T.M., and Harvey, M.E. 1980. Travel trends and energy. In *Proceedings of the 1980 National Outdoor Recreation Trends Symposium*, pp. 59–70. USDA Forest Service General Technical report NE-57. Broomall, PA: Northeast Forest Experiment Station.

Crowe, R.B., McKay, G.A., and Baker, W.M. 1977. *The Tourism and Outdoor Recreation Climate of Ontario*. [Three volumes.] Toronto, Canada: Atmospheric Environment Canada.

Fesenmaier, Daniel R., and Lieber, Stanley R. 1985. Spatial structure and behavior response in outdoor recreation participation. *Geografiska Annaler* Series B 67.

Frazier, J.W., Epstein, B.J., Bell, T.L., and Hillsman, E.L., eds. 1987. *Papers and Proceedings of Applied Geography Conferences, Volume 10*. Binghamton, NY: SUNY-Binghamton.

Goodrich, J.N. 1978. The relationship between preferences for and perceptions of vacation destinations: applications of a choice model. *Journal of Travel Research* 17:8–13.

Heatwole, Charles A., and West, Niels C. 1980. Mass transit and beach access in New York City. *Geographical Review* 70(2):210–217.

Heatwole, Charles A., and West, Niels C. 1982. Recreational-boating patterns and water surface zoning. *Geographical Review* 72:304–311.

Hecock, Richard D. 1970. Recreation behavior pattern as related to site characteristics of beaches. *Journal of Leisure Research* 2:237–250.

Janiskee, Bob. 1980. South Carolina's harvest festivals; rural delights for day tripping urbanites. *Journal of Cultural Geography* 1:96–104.

Janiskee, Robert L., and Lovingood, Paul E., Jr. 1986. Managerial implications of some recent trends in the family camping industry. In *Papers and Proceedings of the 9th Annual Applied Geography Conference*, pp. 154–160. West Point, NY: U.S. Military Academy.

Janiskee, Robert L., and Lovingood, Paul E., Jr. 1987. Campground towns of the grand strand. In R.B. Platt, S.G. Pelczarski, and B.K.R. Burbank (eds.), *Cities on the Beach: Management Issues of Developed Coastal Barriers*, pp. 121–130. Chicago: University of Chicago, Department of Geography Research Paper No. 224.

Knopf, Richard C., and Lime, David W. 1984. *A Recreation Manager's Guide to Understanding River Use and Rivers*. USDA Forest Service General Technical Paper WO-38. Washington, D.C.

Leopold, Luna B. 1969. Landscape esthetics. *Natural History* 78:36–45.

Lieber, S.R., and Fesenmaier, D.R. 1984. Modeling recreation choice: a case study of management alternatives in Chicago. *Regional Studies* 18:31–43.

Lime, David W., and Field, Donald R. 1981. *Some Recent Products of River Recreation Research*. USDA Forest Service General Report NC-62. St. Paul, MN: North Central Forest Experiment Station.

Lovingood, Paul E., Jr., and Mitchell, Lisle S. 1978. The structure of public and private recreation systems: Columbia, South Carolina. *Journal of Leisure Research* 10(1):21–36.

Matley, Ian. 1976. *The Geography of International Tourism*. Washington, D.C.: Association of American Geographers.

McBoyle, Geoff, Wall, Geoff, Harrison, R., Kinnaird, V., and Quinlan, C. 1986. Recreation and climatic change: a Canadian case study. *Ontario Geography* 28:51–68.

McBoyle, Geoff, and Wall, Geoff. 1987. The impact of CO_2-induced warming on downhill skiing in the Laurentians. *Cahiers de Geographie du Quebec* 31(82):39–50.

McMurry, E.C., and Davis, C.M. 1954. Recreational geography. In P.E. James and C.F. Jones (eds.), pp. 251–255. *American Geography: Inventory and Prospect*, Syracuse, NY: Syracuse University Press.

Mitchell, Lisle S. 1981. Fifty years of recreation research. Paper presented at the 77th annual meeting of the Association of American Geographers, Los Angeles.

Mitchell, Lisle S. 1984. Tourism research in the United States: a geographic perspective. *GeoJournal* 9(1):5–15.

Mitchell, Lisle S. 1985. The Tao of recreational geography. *Journal of Cultural Geography* 6(1):141–150.

Mitchell, Lisle S. 1987. Research on the geography of tourism. In J.R. Brent Ritchie and Charles R. Goeldner (eds.), *Travel, Tourism, and Hospitality Research: A Handbook for Managers and Researchers*, pp. 191–202. New York: John Wiley and Sons.

Mitchell, Lisle S., and Lovingood, Paul E., Jr. 1976. Public urban recreation: an investigation of spatial relationships. *Journal of Leisure Research* 8:6–20.

Mitchell, Lisle S., and Smith, Richard V. 1985. Recreational geography: inventory and prospect. *Professional Geographer* 37(1):6–14.

Murphy, Peter E. 1985. *Tourism: A Community Approach*. New York: Methuen, Inc.

Murphy, Peter E. 1987. Commentary on the status of recreation geography in Canada. RTS Specialty Group Panel Discussion Session on 'Mega-Trends in Recreational Geography.' Annual meeting of the Association of American Geographers, Portland, OR.

Nichols, Leland. 1980. Regional tourism development in third world America: a proposed model for Appalachia. In D. Hawkins, E. Shafer, and J. Rovelstad (eds.), *Tourism Planning and Development Issues*, pp. 283–294. Washington, D.C.: George Washington University.

Parsons, J.J. 1973. Southward to the sun: the impact of mass tourism on the coast of Spain. *Association of Pacific Coast Geographers Yearbook* 35:129–146.

President's Commission on Americans Outdoors. 1987. *Americans Outdoors: The Legacy, the Challenge*. Washington, DC: President's Commission on Americans Outdoors.

Ragatz, Richard L. 1974. *Recreational Properties – An Analysis of the Markets for Privately Owned Recreational Lots and Leisure Homes*. Eugene, OR: Richard L. Ragatz Associates, Inc.

Raitz, Karl B. 1987. Perception of sport landscapes and gratification in the sport experience. *Sport Place* 1(1):4–19.

Rooney, John. 1980. *The Recruiting Game*. Lincoln, NE: University of Nebraska Press.

Rooney, John. 1987. Commentary on the status of sports geography. RTS Specialty Group Panel Discussion Session on 'Mega-Trends in Recreational Geography. Annual meeting of the Association of American Geographers, Portland, OR.

Schoolmaster, F. Andrew. 1986. Bank- and float-angler perceptions of use levels on the Madison river, Montana. *North American Journal of Fisheries Management* 6(3):430–438.

Schoolmaster, F. Andrew. 1987. From the chairperson. *RTS Newsletter* (July):1–2.

Schoolmaster, F. Andrew, and Frazier, John W. 1985. An analysis of angler preferences for fishery management strategies. *Leisure Sciences* 7(3):321–342.

Schoolmaster, F. Andrew, and Nelson, Thomas A. n.d. [1984] *Flower Mound Park and Recreation Study: Phase II – Citizen Survey*. Denton, TX: Institute of Applied Sciences, North Texas State University.

Smith, Richard V. 1987a. Applied geography studies. *RTS Newsletter* (July):2.

Smith, Richard V. 1987b. Commentary on the status of applied recreation geography. RTS Specialty Group Panel Discussion Session on 'Mega-Trends in Recreational Geography.' Annual Meeting of the Association of American Geographers, Portland, OR.

Smith, Stephen L.J. 1982. Reflections on the development of geographic research in recreation: Hey Buddy, Can You S'paradigm? *Ontario Geography* 19:5–24.

Smith, Stephen L.J. 1983. *Recreation Geography*. New York: Longman Group Limited.

Smith, Stephen J.L., and Smale, B.J.A. 1980. Classification of vsitors to agreements for recreation and conservation sites. *Contact* 12:35–55.

Stankey, George H. 1973. *Visitor Perceptions of Wilderness Carrying Capacity*. USDA Forest Service Research Paper INT-142. Ogden, UT: Intermontane Forest and Range

Experiment Station.

Stankey, George H., and Baden, John. 1977. *Rationing Wilderness Use: Methods, Problems, and Guidelines*. USDA Forest Service Research paper INT-192. Ogden, UT: Intermontane Forest and Range Experiment Station.

Stansfield, Charles A. The geography of resorts: problems and potentials. *Professional Geographer* 23(2):164–166.

Van Doren, Carlton, S. 1967. A recreational travel model for predicting campers at Michigan state parks. Unpublished Ph.D. dissertation, Department of Geography, Michigan State University, East Lansing, Michigan.

Van Doren, Carlton S. 1975. Spatiality and planning for recreation. In B. van der Smissen (compiler), *Indicators of Change in the Recreation Environment*, pp. 335–358. Penn State HPER Series 6. College Park, PA: Department of Health, Physical Education, and Recreation, The Pennsylvania State University.

Van Doren, Carlton, S., ed. 1984. *Statistics on Outdoor Recreation – Part II, The Record Since 1956*. Washington, DC: Resources for the Future, Inc.

Wall, G., Harrison, R., Kinnaird, V., McBoyle, G., and Quinlan, C. 1986a. Climatic change and recreation resources: the future of Ontario wetlands? *Proceedings of the Applied Geography Conferences* 9:124–131.

Wall, G., Harrison, R., Kinnaird, V., McBoyle, G., and Quinlan, C. 1986b. The implications of climatic change for camping in Ontario. *Recreation Research Review* 13 (1):50–60.

West, Neils A., and Heatwole, Charles. 1979. Urban beach use: ethnic background and socio-environmental attitudes. In Proceedings of the Fifth Annual Conference of the Coastal Society, *Resource Allocation Issues in the Coastal Environment*, pp. 195–204. Arlington, VA: Coastal Society.

Wolfe, Roy I. 1964. Perspective on outdoor recreation: a bibliographic survey. *Geographical Review* 54: 203–238.

Wolfe, Roy I. 1967. A theory of recreational highway traffic. DHO Report RR129. Downsview, Canada: Department of Highways of Ontario.

Wolfe, Roy I. 1978. Vacation homes as social indicators: observations from Canadian census data. *Leisure Sciences* 1: 327–343.

Wolfe, Roy I. 1980. Pattern of recreational highway traffic in time of energy scarcity. In *Contemporary Leisure Research, Proceedings of the Second Canadian Congress on Leisure Research*, pp. 503–507. Toronto, Canada: Ontario Research Council on Leisure.

Robert L. Janiskee and Lisle S. Mitchell
Department of Geography
University of South Carolina
Columbia, SC 29208
U.S.A.

11. Working Both Sides of the Street: Academic and Business

During the past ten years the applied side of geography has come to occupy an increasing amount of the discipline's attention. Stimulated by a decline in academic job opportunities, applied geography has become an attractive alternative to university teaching and research. As with any developing field of inquiry, its form, shape, structure, and direction are not well defined. Some suggest that applied geography applies geographical techniques to the solution of practical problems (Frazier and Henry 1978, p. 4). Others, not in disagreement, add a social dimension to the definition, indicating a need to evaluate human welfare (Frazier 1982, p. 13). To some, applied geography centers on ends rather than means, emphasizing results rather than methodology (Sant 1982, p. 1). A less restricted view considers that applied geography is the use of geographical knowledge as an aid to making choices (Briggs 1981, pp. 1–8). As expected, there is a divergence of views on how to define applied geography. Two underlying factors, however, appear common to all definitions of applied geography: it takes place outside the university, and it deals with real world problems.

Attempting to sort out definitions and directions in applied geography is frustrating, because there is little common ground upon which to stand. It becomes apparent that without a commonly agreed upon definition of geography, applied geography then assumes the same amorphous shape as does geography. Thus, to identify and evaluate what is 'academic geography' and what is 'applied geography' is almost impossible. Rather than write another essay on the definition of applied geography, or one extolling the need for more applied geography, or compose yet another example of how to do applied geography, this essay will reflect upon my own background as an applied geographer of sorts for the past fifteen years. My views are somewhat different and personal about the nature of applied geography and I will not attempt to complicate matters or bore the reader with lengthy justifications.

This essay relates my experiences as a teacher of applied geography, as a professor of historical geography, and as president of an eight-employee consulting firm. Each role has provided me with a different view of geography's ability to solve problems. My experience suggests that applied and academic geography are identical in subject matter and barely distinguishable by what each attempts to achieve.

Teaching Applied Geography

Applied geography has become very much in vogue during the past five years,

M. S. Kenzer (ed.), Applied Geography: Issues, Questions, and Concerns, 165–172.
© 1989 *Kluwer Academic Publishers.*

particularly in planning for new curricula. With declining student enrollments in geography, there has been an emphasis on career development as a means to boost enrollments. According to this scheme, a student in anticipating a job after graduation will major in geography only if the curriculum can be revised to illustrate that there are specific jobs for geography graduates. Many departments of geography have adopted this strategy and some have found success. However, my experience in developing and maintaining an applied geography program during the past fourteen years has been that it is not curriculum changes that are required to have a first-rate applied program, but establishing instead a sound career counseling program for geography majors in lieu of specific career oriented courses.

Modifying a curriculum to include a battery of new courses that emphasize technical skills for a job after graduation is at best a temporary and very limited solution to enrollment problems. Although this solution would probably increase the number of employed geography graduates, efforts to adopt applied curricula in response to employment needs may well lead to short-term gains and long-term losses. With a concerted effort to train students for jobs we may focus our attention too narrowly, thus developing a restricted view of teaching and geography. Any drastic alteration of curricula and teaching methods would not be in the best interest of our students. The adaptation of courses and teaching methods to meet the continuing needs and fluctuation of the job market is impossible. The job market is constantly changing and I doubt that any geography department has the money, the staff, and the equipment to keep up with continual change.

The solution is not to make drastic changes to the curriculum, but to create instead a career counseling program for geography majors. The inability of geography majors to acquire employment stems in many cases from the students' inability to recognize his or her skills, to market them properly, to plan for employment after graduation, and, importantly, to obtain career counseling early. A major goal of such counseling is to help students identify their abilities and to convince them that after graduation they will possess marketable skills. It is also important to train the faculty to think in terms of career advising; a successful applied geography program depends in large measure on informed faculty willing to discuss career strategies with students.

Most geography curricula attempt to provide students with a broad background as opposed to professional and vocational programs that furnish specific techniques and programmed routines applicable to a handful of selected jobs. To some, this liberal orientation might suggest that geographers are trained generalists, and that there seems to be little demand for generalists in the current job market. New geography graduates, however, are not entirely devoid of technical skills. Although the discipline in most universities and colleges is less technical than engineering or computer science, geography does provide statistical, cartographic, and computer application skills. However, the popular notion that employers are seeking graduates with technical backgrounds and specialized skills is not true for all jobs or all employers. From my experience in assisting students during the past fourteen years in applying for jobs it is a fact that, by far and away, employers are seeking a well educated and informed graduate.

A few of the skills acquired by the well educated geographer, regardless of the program from which he or she graduated, include understanding written material, writing effectively, identifying problems and proposing practical solutions, and performing detailed and accurate work. These skills may seem commonplace and hardly worth trying to sell in a highly competitive marketplace, but to an employer they are, in fact, scarce skills. The technical skills acquired from applied geography programs are really short-term skills that will change with the marketplace, but the ability to identify problems and to propose solutions, to communicate, and to organize data will not become outdated with the next generation of computers.

During the past three years I have been the Director of Business Research at California State University-Northridge, and I have been responsible for developing and maintaining research contracts with the business community. In addition, I have established a paid, student intern program in which students from the social sciences and the business school work on projects for local business clients. After only a few weeks into a semester, students are quickly involved with solving very practical business problems. Students are taken to business meetings, and then they have to prepare proposals and negotiate fees. Ultimately, they become responsible for completing the project on time, within budget, and making the final presentation to the client. Recent projects have included preparing a simple business plan for a small business, creating marketing strategies, conducting market surveys, organizing a computerized database, modeling business growth, and developing long-range business plans.

It is not uncommon to find marketing and management majors working with geographers or historians. It might be expected that in a business situation business majors have the advantage, but this is not usually the case. A well educated geographer inevitably has better problem solving skills, better communication skills, and definitely better presentation skills than most business majors. I might add that most of the other social science students working as business interns also demonstrate excellent problem solving and communication skills. It might be suggested that I am biased in favor of geography and the social sciences in general, but let me caution that in dealing with the business community I cannot afford to play favorites. The Bureau of Business Research is self supporting, and without continued support from the business community it would not exist.

From my experience as a career counselor for geography majors and even my association with the business community, I can say that geographers need not rush wholesale into developing applied geography courses to achieve jobs for their majors, but should consider strengthening their core curricula to include more breadth in the discipline and emphasize analytical and communication abilities. In addition, geography departments should try to develop career counseling programs, if for no other reason than to give geography majors self confidence and the ability to examine their own potential.

Academic Applied Geography

My consulting experiences outside the university are not, I am sure, much different
than those of other university professors. As a historical geographer, I have worked
for attorneys as a legal research analyst, acted as an expert witness, consulted on
archaeology projects, assisted in preservation efforts, written articles for
newspapers, coordinated museum exhibits, and delivered a large number of lectures
to community groups. While I was paid for my services, I have never thought of
these activities as applied geography, but rather as an extension of my academic
research interests in the Spanish settlement of the American Southwest. My
academic interests, therefore, have led to working with individuals who need the
special knowledge I have gained through my research and are willing to pay for my
time.

As a historical geographer I am delighted every time I am asked to work on a
project outside the university, not because of the applied aspects or for the consulting fee, but because I am educating a broader, professional audience about
historical geography and geography in general. This type of applied geography is
not based so much on techniques or application, but on specialized knowledge.
What is interesting is the variety of questions asked. Most are very specific and
usually require detailed answers that depend more on judgement than on an
elaborate set of methodologies or techniques.

Some examples of my recent historical geography consulting projects include the
following: working as an expert witness dealing with questions of Spanish and
Mexican land tenure in California, specifically considering whether the boundaries
of a Mexican rancho extended to low tide along San Francisco Bay; asked to
explain how Mexican land boundaries were established and maintained; and asked
how land use patterns under rancho settlement affected erosion and subsequent
drainage patterns along the southern California coast, to determine whether an
island existed before 1850. As a cultural geographer, I have been asked to assess
and evaluate the authenticity of Hispanic buildings for preservation efforts, and to
investigate the development of historical irrigation systems for a specific region
and recommend preservation action.

While the above list may, at first glance, appear to be very broad, the projects all
have two things in common: each relates in some way to Hispanic settlement of
California, and each was a very specific question about a specific topic. I was asked
to participate not because I was considered a historical or applied geographer, but
because I had special knowledge about the subject and could answer, with some
degree of authority, the questions posed. Thus, the actual item sought was my
competency.

From my perspective, geography has very little to do with me acting as a
consultant on historical matters; indeed, a historian, anthropologist, or an engineer
would suffice if they had the in-depth knowledge sought. A great deal of this type
of consulting is not amenable to publication in academic journals. Usually the
questions are very narrow and place or time specific; the techniques employed are
often simple, and usually the research documents are well known. Frequently, the

person acting as a consultant has already addressed the question in previous research, and that is why he or she is solicited to work on the project. Therefore, recognition by one's colleagues for conducting this type of research is not frequent.

How does academic consulting fit into applied geography? The answer is that it depends entirely on how applied geography is defined. If applied geography is defined as the solution of practical problems, then academic consulting falls within the definition, and many university faculty members would qualify as applied geographers. Academic consulting is an extension of one's ongoing research, and the procedures, techniques, and methods are academic in nature. The only difference between what is usually identified as applied geography and academic consulting is how one defines a practical problem.

Business Consulting

As the president of a location and marketing consulting firm, my views on applied geography are substantially different from my academic and teaching interests. My firm employs eight geographers who range from recent B.A.s to Ph.D.s. We provide site evaluation, location analysis, marketing research, and custom computer cartography to financial institutions. Examples of some of the projects completed recently include a site evaluation of over 7,000 banks in California for a client who wanted to establish a new marketing strategy based on competitor location and site characteristics, a reconfiguration of a branch network including closing current locations and recommending new sites, a complete market analysis on the Florida banking market, a plan to develop a segmentation scheme for market penetration, and the identification and evaluation of growth communities in California. As geographers, we used maps as a simple means to convey results; today, however, preparing custom maps for our clients has become a major part of the business.

Working in the business world as a geographer is very different than teaching applied geography or academic consulting. Working the other side of the fence brings to view a set of separate and distinct problems not usually found in the academic world or contained in the latest applied geography missive. Clients have very specific problems to be solved; they neither have the staff nor the expertise to solve the problem themselves and thus require the services of a consultant. Unlike academic consulting, where expertise in a relatively small area is required and solutions are often based on academic judgement and standard methodologies, business requires specific solutions to specific problems – problems most frequently that are unique to the client. Thus, there is no standard solution and no set procedure or methodology to follow. Yet, our clients need practical solutions to their problems: solutions that will be accepted by upper management, that can be implemented with existing staff and budgets, and that provide cost effective results. In addition, our clients always have very specific deadlines that have to be met without fail. The best solution to a business problem is worthless unless it can be delivered on time, within budget, and implemented efficiently within a corporate structure. The experiences gained from teaching applied geography or those developed from academic consulting did little to prepare me for the practical world of business.

Business geography is at once rewarding and frustrating. It is frustrating because the clients' problems are often so amorphous that they know neither how to state them properly nor how to approach a potential way to solve them. All the client knows is that he or she has a problem. Our first task, therefore, is to work with them to define the specific problem, pose specific questions to be addressed, and then propose various ways to solve the problem. Solving the problem can be exciting because it is intellectually stimulating. Yet, any solution must ultimately be practical, understandable, and cost effective. Thus, solutions that are proposed are not always the most challenging, the most innovative, or the most creative. Solutions must be usable by the client. If, for example, a client does not understand statistics, then to solve the problem statistically would be of no value to that client.

The application of geographic techniques and principles to the solution of business problems is infrequent. Unlike the occasional article that documents a specific business problem solved by a geographer in a geographic manner, the norm is to apply practical knowledge and common sense. A site evaluation accomplished by textbook methods, for instance, will be accurate and geographic but frequently cost-prohibitive to the client.

Many geographers have tried to establish site location firms, but they quickly learn that they have little to offer that is unique. Many firms with multiple locations have survived and have made a profit without the services of a geographer. A branch network is often developed by one of three methods: guessing (commonly known as common sense); a study by someone in the firm (often not a geographer); or a study by an outside consultant (almost always not a geographer). Since many businesses have been successful – those not successful are no longer in business – in locating sites without the services of a geographer, how does one convince a potential client that there are specific geographic principles and techniques that would provide better analysis and yield better sites? The task is difficult, to say the least, because business does not understand geography, has never heard of applied geography, and probably never will. Thus, from a business point of view, what could a consulting geographer possibly bring to a firm with a successful branch network established by using 'common sense'? The answer is simple: solutions that fit business's needs, not geographical solutions.

Business geography, then, is not really geography but business. My consulting firm depends not on the application of geographical techniques to the solution of practical problems, but on sound business practices, a well trained staff (who all happen to be geographers), product development, anticipation of future market requirements, and the ability to respond quickly to a client's needs in a cost-effective manner. The reason for the business is not to convince clients of the rightness of geography, but to sell workable solutions for a profit.

From a business perspective, my view of applied geography is that it has very little to offer in the way of practical solutions to my clients' needs. Published applied research often differs little from academic research, except that there is an attempt to show some practicality. The same techniques, methodologies, and arguments are used as in academic publications. It has always appeared to me that applied geography is conducted as a research grant for a government agency, and most of it is. If the applied geographer is to survive in the business world, he or she must quickly shed their academic leanings, learn to understand budgets, costs, and time limitations, and develop a common sense approach to problem solving within a client's needs. In addition, much of the research conducted for business is proprietary, thus there are few opportunities for a publication to enhance the résumé; those who want academic rewards should not seek it in the business world. There are two rewards for the business geographer: the clients pay their bills, and they ask to have another problem solved. I doubt that much of business geography will ever grace the pages of the *Annals* for two reasons: it might not withstand the academic review process, because it might not be up to academic standards, and the problem itself probably would be of little interest to the readers.

Conclusion

I began this essay with a brief review of applied geography definitions, and I am inclined now to think that there may not be an all inclusive definition. I do think, however, that much of what is labeled 'applied geography' differs little from 'academic geography,' or 'academic consulting,' except that applied geography appears to be practiced by those not in a university and is heavily laced with techniques. Moreover, I do see a great deal of difference between business geography and applied geography. Applied geography is much too academic, appears to be conducted as a research grant, and always solves a major 'practical problem.' Business geography, on the other hand, has few academic trappings, is less pretentious in its findings, and necessarily contains a great deal of common sense. To be fair, however, the recent surge in applied geography is an attempt to develop a meaningful view of what geographers can do outside the university. Business geography, on the other hand, is profit motivated, with little or no concern with promoting geography as a discipline or as meaningful experience.

The most difficult aspect of applied geography comes from those within the university who look upon applied geography as a savior of the discipline. They preach applied geography and want a curriculum to match their enthusiasm. In their zealousness, they may shortchange the students by giving them a restricted view of employment opportunities after graduation, suggesting that they are limited in the type of jobs available to them. While technical skills are important, a solid geographical education will provide the student with even greater career opportunities. The push to modify the curriculum to include technical training says a great deal about geography and geographers, particularly that we have no vision about what it is and what it stands for. Simply stated, we lack faith in ourselves as geographers. If we shift and shuffle to every popular whim, and if we realign geography to conform to university enrollment patterns, then we surely risk becoming a conformist discipline without a core, constantly defining and redefining ourselves to the public and to academic administrators.

References

Briggs, David 1981. The principles and practice of applied geography. *Applied Geography* 1:1–8.

Frazier, John W., editor, 1982. *Applied Geography: Selected Perspectives*, New Jersey: Prentice-Hall.

Frazier, John W. and Norah F. Henry 1978. Selected themes in applied geography. *Geographical Survey* 7:19–26.

Sant, Morgan 1982. *Applied Geography: Practice, Problems and Prospects*, London: Longman.

David Hornbeck
Department of Geography
California State University
Northridge, CA 91330
U.S.A.

The 'Taken for Granted' Side of Applied Geography

12. A Critical Appraisal of 'Applied' Cartography

Maps are nothing if not useful. Some communicate information in passive, digital form, while others beam their messages with great visual impact. Maps can also confuse, mislead, entrap, or persuade insidiously. But all maps appeal to the human mind because they represent places, and because they are *applicable* to the universal need for place-related knowledge. Geographers accept maps as indispensable *tools* for the identification, analysis, and solution of spatial problems. In fact, this direct link between geographers and maps may be the public's most dominant image of geography.

It is natural that academic cartography be conceived as an 'applied' field in which much time and effort is devoted to applying particular cartographic techniques to specific geographic problems. This conception is nearly universal and possesses an historic lineage. Yet formal attempts to define the field on an *applied* basis have not been fruitful. In an early paper, Eckert ([1908] 1977) considered cartography an 'applied art.' Later, Imhof ([1963] 1977) used the phrase 'applied science.' Ratajski (1973) tried to place applied cartography within an elaborate theoretical structure for the whole of cartography, but admitted that it had no theoretical foundation of its own. There has also been an attempt to link the study of the history of cartography to applied cartography (Woodward 1974). Generally, attempts to define the applied nature of cartography have met with little success, and have not clarified the field's pedagogical and research agendas.[1]

Cartographers have been concerned with two principle traditions: the intersections of maps and technology, and those of maps and humans. Cartography is probably the most applied subfield of geography because theoretical work in the first tradition is rare, and that in the second has yet to bear much fruit. Yet theoretical inquiries in cartography have nearly disappeared; little has appeared since *The Nature of Maps* (Robinson and Petchenik 1976). An operational view of cartography has dominated since the mid-1970s, perhaps in reaction to *The Nature of Maps*. The one exception – an important text written by British cartographer, J.S. Keates (1982) – is apparently consulted quite infrequently judging by cartography journal article references. In fact, introspection and general musing over the persistent theoretical problems in understanding maps, humans, and technology seemingly has disappeared from our thoughts.

If attempts to define cartography as an applied field have been unsuccessful, yet there persists a studious avoidance of basic, theoretical research with potential lasting value, it would appear that either cartographers are incapable of shaping their own subdiscipline, or are devolving to the status of technical consultants to other fields. In this chapter I provide a critique of the current emphasis on application-oriented cartography and make a plea for redirecting *some* of our

M. S. Kenzer (ed.), Applied Geography: Issues, Questions, and Concerns, 175–191.
© 1989 Kluwer Academic Publishers.

efforts back to basic research and education. I believe it is important to place the current situation in a historical context in order to understand persistent patterns in the ebb and flow of the history of academic cartography. Thus, I will provide first a brief review of cartography's role in earlier periods when applied geography was popular, followed by an assessment of the short- and long-term impacts (on students, on our relationship to both private- and public-sector employers – 'the industry' – and on our geographic colleagues) of a renewed interest in an application-oriented cartography. Three themes emerge: since World War II cartographic research has been guided by the mind-as-machine metaphor that swept all of the social sciences; the meaning of applied cartography is ambiguous through time; applied cartography drives or catalyzes periodic applied geography 'movements.'

Antecedents of Contemporary American Academic Cartography

Cartography as Mapmaking (1895–1938)

Since the earliest days of American geography, cartographers were recognized as a separate breed of professional, nonacademic geographer. No professional cartography organization existed, nor was a formal subspecialty to develop for several decades. Nevertheless, cartographers were among the first geographers to be recognized as a distinct group. The most influential American geographer at the turn of the century, William Morris Davis (1850–1934), delineated five classes of 'professional geographers': teachers, writers on geographic subjects, explorers, topographers, and cartographers (Dunbar 1981, p. 77). Although we are unsure, we must assume that Davis saw cartographers as mapmakers, not as teachers or writers on cartographic topics. Such an application-oriented vision surely would have dominated early thinking in academic geography. There was little interest to develop a theory of cartography, nor was there concern for map communication, accuracy, and the psychological characteristics of the typical map reader, which were all to emerge later.

World War I had obvious implications for cartography. Numerous geographers were called to service, and many made maps or taught mapmaking, mapreading, and surveying at home or abroad. For example, Lawrence Martin (1880–1955) trained students in military mapmaking at the University of Wisconsin in 1917. Others served on the American Commission to Negotiate Peace in Paris during the immediate postwar period (1918–1919): Mark Jefferson (1863–1949) was chief of the cartography division for the Commission; Bailey Willis (1857–1949) directed the preparation of the new 1:2,000,000 map of South America; Armin K. Lobeck (1886–1958) produced maps of politically sensitive areas (e.g., the Balkan States); W.L.G. Joerg (1885–1952) developed a map of Germany's new boundaries, which was released with the draft of the peace proposals on May 7, 1919.

The war had a significant impact on cartography for many postwar years. Most important may have been its growth in stature, both within and outside geography.

The Western world realized that gaps existed on the globe where relevant mapping was either inadequate or nonexistent. In addition, nineteenth century tools and mapmaking methods were recognized as ill-suited to postwar mapping needs. As geographers returned to the workplace, interest in applied geography accelerated, and cartography played a major role in its development.

In the period immediately following World War I, cartographically inclined geographers, and others with an interest in maps, focused on the development and application of specific types of maps and map symbols, and on the methods of making them. Superior innovations were technical, based on trial-and-error, intuition, and a critical judgment gained from experience. American geographers were more interested in applying map techniques to specific problems than in the theoretical aspects of cartography; wars rarely produce a concern for theory, as the needs for applications are too pressing.

In the 1920s and early 1930s, when the distinction between cartographers and geographers was blurred, achievements were numerous. Among these were large-scale land use/land capability mapping techniques (Barnes 1929; Finch 1933; Hudson 1936; Jones and Finch 1925; Sauer 1919), new projections (Goode 1929), statistical mapping of agriculture and population (Baker 1921, 1926; Jones 1930; Smith 1928; Wright 1936), methods of surveying and mapping using aerial photographs (Miller 1931), and the use of map analysis to resolve boundary disputes (Martin 1930).

The few who called themselves 'cartographers' were concerned almost exclusively to develop methods for displaying the form and aspect of terrain features. Lobeck, Francois E. Matthes (1874–1948), and Guy-Harold Smith (1895–1976) were the most important in this era. Terrain maps from this period generally appear as illustrative and explanatory graphics in books (e.g., Fenneman 1931; Lobeck 1933), or as separate map sheets published by regional, state, or federal agencies. One notable exception is Lobeck's (1924) text, which was devoted entirely to the production of terrain maps.

The first era of American academic cartography culminated with Erwin Raisz's (1893–1968) *General Cartography* in 1938. Not only was this the first comprehensive American cartography text, but it was also the first book to demonstrate known mapping techniques and their application to geographic data. It is an applied *tour de force* with many nuggets of information still useful today, but it contributed little to cartographic theory.

Another War and a New Cartography

The 1940s brought World War II and a resurgence in applied geography. The war's impact on cartography, including the emergence of widespread government influence, the formation of professional organizations, and concern over the breadth and quality of cartographic education, tugged and pushed the new specialty in different directions.

In 1949, the federal Office of Naval Research (ONR) began to offer contracts for geographic research, a source of funding that emergent cartography specialists would tap for many years (Association of American Geographers 1949). Between 1949 and 1953, more than 200 geographers were given ONR assistance. The ONR recognized that basic research in geography and cartography was a pressing postwar need, and they indicated that proposals for both applied and theoretical research would be considered equally, that no particular topic would be favored. Their general announcements in geography periodicals stressed the removal of obstacles to this goal caused by security regulations and restrictions on types of research funded (Association of American Geographers 1953). Presumably, all topics would be entertained and reviewed by an advisory committee that included many academic geographers. This was not always the view implicit in calls for proposals made by individual ONR representatives:

It is hoped that a number of geographers interested in academic research ... will communicate with the Office of Naval Research, outlining their proposed studies so that research projects can be integrated and *suggestions can be made to bring their work more closely into line with the needs of the armed forces* [emphasis added] (Ridge 1948, p. 20).

In a special report issued by the American Society for Professional Geographer's (ASPG) national Committee on Cartography, college-level cartographic coursework was specifically outlined (Quam 1946). Lower-level training in map appreciation, reading, interpretation, and design would build toward upper-division work in 'professional cartography.' A checklist of objectives and topics emphasized technical competence and familiarity with the type of techniques, materials, and tools used by mapmakers in the federal government. The objective was to *train* students (in the manner and conventions already in use) for readily available government jobs. The idea of 'university-as-vocational-school' was not new, but for the first time academic cartographers and geographers were asked to adjust their course content to suit the perceived needs of military and intelligence sectors of the federal bureaucracy. This should not have been surprising since the 'national' committee consisted entirely of representatives from various military and intelligence branches of the government, attesting to the ASPG's ability to recruit government bureaucrats and technocrats into its membership.

The military's incursion into university-level cartographic instruction was remarkable. Faculty were recruited and retained, and courses were taught, with the idea that a cheap labor force was awaiting proper training before entry into the field of military mapping. In 1951, the Army Map Service (AMS) Program in Applied Cartography (directed by Frances Mae Hanson, an associate professor at the University of Pittsburgh) began developing courses in 'applied cartography' and 'map intelligence' in colleges in the eastern United States. By 1952, twenty-five institutions of higher learning had enlisted in the program (Hanson 1952). The courses used standardized visual and textual aids, and emphasized practical training in technology used by the AMS to enable student recruitment into service without delay should the need arise.

Florida State University was perhaps a typical AMS training center. In 1951, they initiated an applied cartography course taught under the auspices of the AMS.

The instructor was AMS-trained, and the university's Geography Department attained 'approval' as a site for army training programs should conflict break out. Here, 'applied cartography' quite clearly meant the application of knowledge about maps and mapping to specific problems *as it benefited the military and intelligence branches of the federal government.*

Whether formally affiliated with the AMS or not, university courses in applied cartography began to sprout across the continent around mid-century: Michigan State University offered an advanced 'applied cartography' course in 1949; Kent State University and the University of Washington were doing the same by 1951; Ohio State University began an interdisciplinary Institute of Geodesy, Photogrammetry, and Cartography in 1951 (the program offered a separate B.S. degree in the mapping sciences). The University of British Columbia offered the first full-year course in cartography in Canada in 1950; McGill University offered 'practical cartography' in 1954.

The Emergence of Professional Organizations and Concern for Cartographic Education

Three professional organizations emerged in the 1940s to represent the burgeoning postwar interest in cartography: the ASPG, the American Congress on Surveying and Mapping (ACSM), and the Cartography Committee of the Association of American Geographers (AAG). Most members of ASPG and ACSM were government cartographers and those in academia favoring continued emphasis on the technical achievement and skills needed for mapmaking in various branches of government. The AAG's Cartography Committee tried to move in another direction, emphasizing cartography as a scholarly endeavor intimately linked to geography.

In the 1940s, the Washington, D.C. mapping community was dominated by 'cartographic engineers' working on topographic maps – not 'geographic cartographers' interested in mapping geographic phenomena (Voskuil 1950). The engineers, or 'cartotechnicians' as Erwin Raisz (1950a) called them, possessed different qualifications and were concerned with production-mode cartography in a 'can-do' work environment where success was measured by the number of maps produced and the technical accomplishments and efficiencies achieved in the mapmaking process. Once the war began, a number of these government cartographers met and formed a professional association. Although no discrimination of geographic cartographers was intended, only three were present at this formative ACSM meeting: W.L.G. Joerg of the National Archives, S.W. Boggs (1889–1954), the Geographer at the State Department, and John K. Wright (1891–1969) of the American Geographical Society (Ristow 1983). The ACSM was officially organized in April, 1942, and although a 'Division of Cartography' was set up by Wright to include such topics as compilation, symbolization, and the history of cartography, from its inception the Congress (and its house publication, the *Bulletin*) was slanted toward topographic engineering and surveying applications of cartography.

In 1947, Raisz sent a letter to AAG members on the possibility of forming a special cartography group within the AAG. On December 30, 1948, at the Madison, Wisconsin AAG meeting, the Committee on Cartography was formally charged with three responsibilities: stimulating cartographic research, developing a special session on cartography at the next annual meeting, and cooperating, via a designated liaison, with the ACSM. The organizers were intent to promote cartography at all levels, but especially on university campuses (Raisz 1950b). Later, the goals were developed more broadly: to raise cartography to a level as an important subfield of geography, and to prepare geographer-cartographers for work and for research (Raisz 1952). Additionally, special cartography issues of the *Professional Geographer* appeared from 1950 to 1953 and again in 1955. From the beginning Raisz drew a line between the purpose of the AAG Committee and the interests of the majority of cartographers in the federal bureaucracy (Raisz 1950a); the AAG was to serve 'geographic cartographers,' not the unscholarly, purely technical concerns of the federal bureaucrats.

The formation of the AAG Committee represented a somewhat daring break from tradition. One outcome of World War II was the recognition of woeful inadequacies in the quality of map production, knowledge about maps, and the number of people with knowledge about maps (Robinson 1952, 1954). Arthur Robinson was selected to head the Cartography Section of the Geography Division of the Office of Strategic Services (OSS) – the forerunner of the Central Intelligence Agency – after only a single course in cartography, yet this represented more experience than most American geographers had in 1941. The need for postwar university training in applied cartography for government jobs appeared unassailable. Raisz's intent to have the AAG group represent the scholarly side of cartography thus appears at odds with these needs. But he was not the only cartographer calling for greater focus on basic or theoretical postwar cartographic research, nor was he the only one to set himself at odds with those in the federal government (e.g., Harrison 1950; Robinson 1954).

The move toward greater theoretical research, and away from rote learning of conventional practice, reached its peak in the immediate postwar period with Robinson's *The Look of Maps* (1952). This book symbolizes four major achievements: the establishment of cartography as a field of scholarly inquiry in the United States; the first book-length analytical approach to cartography; the first systematic analysis of flaws in standard government practice; the establishment of a role for psychological research in the development of cartographic theory – the signal call for experimental cartography.

Concern over the breadth and quality of university training in cartography emerged in the 1950s (Beishlag 1951; Jenks 1953; Mackay 1954; Odell 1950; Robinson 1951). Knowledge of fundamentals, especially processes of selection, classification, and generalization, and a broad understanding of geography were seen as indispensable parts of a cartographic education. Less attention was devoted to standardized instruction or special familiarity with government mapping materials and procedures. In fact, university-level pedagogic trends eventually veered away from AMS-like sponsored programs. In an attempt to unite academic cartography with geography, rather than with government needs, Philbrick (1953) suggested that solutions to specific mapping problems lie in the skillful application of general cartographic principles, not in a lengthy checklist of memorized rules.

A tone had been set by the end of the 1950s: government agencies and private companies continued to demand applied cartographers and they encouraged applied research in the universities, whereas academic geographer-cartographers developed interests in basic research, using the paradigm of psychophysics. Rapidly accelerating technological change and the need for military preparedness provided academic cartography with a strong applied cast, but research cartographers were entering a period when theoretical concerns were gaining prominence for the first time. This era has been reviewed thoroughly already (Castner 1983; Petchenik 1983), but it is worth noting that many academic cartographers now view this period as an ultimately unsuccessful attempt at theory development. For this and other reasons, the applied view reappeared in the mid-1970s.

The Current Conceptualization of Maps

By the 1970s, we began to regard 'information' and 'knowledge' as primarily technical achievements based on transmission efficiencies in the accumulation of facts that could be bought and sold; now, we increasingly measure American society's progress in such technical terms rather than cultural ones (Roszak 1986). Maps are currently considered bearers of 'information' feeding an emergent 'information society,' and cartographers have become 'spatial information specialists' concerned with technical accomplishments in the storage, retrieval, and manipulation of information in machines, not with creating and interpreting maps.[2]

Today's mechanistic view of maps and the human mind encourages a wholly application-oriented approach to cartography at a time when other factors have stimulated a resurgence of interest in applied geography. Disappointments in attempts at high-level theory production, concern for 'relevance' to so-called real world problems, the spiralling decline of student enrollments and concomitant threats of department closure, increasing differences between faculty incomes and living costs, another postwar era of technological change, the supremacy of the 'publish or perish' mandate, and the contemporary student's vocational approach to higher education (in direct opposition to the rampant idealism of the late 1960s and early 1970s) were instrumental in reviving applied geography.

In cartography, events paralleled, and seemingly catalyzed, these general trends in geography. Dramatic changes in technology – especially the advent of desktop microcomputers, powerful minicomputer systems, and output devices capable of ever-finer resolutions – and the perceived failure of the experimental paradigm to develop cartographic theory were important factors in turning cartographers once again toward technical problem solving. Government- and private-sector cartographers were ill-disposed toward theory development and encouraged the applied alternative sought by disenchanted cartographers (cf. Robinson and Petchenik

1976; Petchenik 1985). The phrase 'applied cartography' is still used, but far less than in the post-World War II era (Dahlberg 1978; Rabenhorst and McDermott 1988; Ratajski 1973).

The current bout with applied cartography is somewhat similar to that of the 1950s. Professional organizations continue to emerge, or old ones recombine to form temporary amalgams to better capture a fickle membership. This activity divides cartographers with specific technological interests into separate, homogeneous groups.

It would be pretentious here to attempt a comprehensive review of current applied research in cartography; such exercises disclose the problems of separating purely applied research from the remainder. But general categories can be established that embrace most, if not all, of this work, by using the chapter titles from Mark S. Monmonier's *Technological Transition in Cartography* (1985): 'Location and Navigation,' 'Boundaries and Surveys,' 'Aerial Reconnaissance and Land Cover Inventories,' and 'Decision Support Systems.' The first has obvious military applications. In the second chapter, Monmonier argues that mapping could be conceived as a publicly owned and operated utility, wherein geodesy and boundary control are maintained by governments to serve the public good (cf. Wellar 1985). The third suggests map content, and includes especially the development and maintenance of digital data bases derived from remotely sensed images. The fourth signifies GIS research, and research on expert systems, AI, and unknown future technologies.

Another view of applied research in cartography is supplied by Degani (1980). He casts a finer net over the myriad possibilities than Monmonier does and develops eight groups based on the *utility* of the map, especially its ability to contribute to the solution of spatial problems: maps displaying relations between different parts of space, especially causal relationships; maps dividing space into functional parts (e.g., political) for the purpose of displaying proximity, accessibility, etc.; maps used in optimum location planning; maps for optimum spatial allocation of services; maps for conducting network analyses, especially for transportation planning; maps assisting determination of the influence of certain spatial characteristics on phenomena; maps depicting simulations of future spatial patterns; and maps identifying the spatial attributes influencing distributions.

Given the generalization (and resultant ambiguities and overlaps) in these categories, it is easy to appreciate the difficulty in defining the contemporary applied movement in cartographic terms. In essence, contemporary applied cartography research is that which is devoted to either military or primarily civilian applications – although the boundary here is increasingly vague in cartography – where the major concern is solution of a specific problem, or development of a technique, without embedding that solution or technique within a broader cartographic theory. This avoids establishing principles or guidelines stemming from research that otherwise might realize wider applications.

A major distinction must be drawn between past and present concerns for cartographic applications. This distinction mirrors the changes in American society as a whole. Beginning about 1980, the U.S. fully embraced the so-called information age and initiated both a military build-up and a 'spend-up' of unprecedented proportions for a peace-time period. These two trends are linked, and it is easy to envision the role cartographers may play in them as the end of the century unfolds.

Presently, a 'cult of information' has developed in which cartographers increasingly seek to become involved. Yet we have exercised little critical thinking in recognizing the pitfalls of this most recent manifestation of applied cartography. In the following section I identify three problem areas: our relationship to students, industry, and our colleagues in geography. Some of my observations are either extracted or adapted from recent critiques of the role of technology in American society (Donnelly 1985; Dreyfus and Dreyfus 1986; Roszak 1986).

Cartographic Training or Education?

'The distinction between education, a process aimed at drawing out the abilities of the student, and training, in which the student is learning to negotiate a structured domain, is crucial' (Dreyfus and Dreyfus 1986, p. 135). Applied cartographic instruction places its emphasis on *training*, not education. The most serious danger facing cartography instructors today is the emphasis on training in the manipulation of 'turnkey' computer systems, to the exclusion of cartographic education. Students see mapping variables made precise and explicit by computers; hence they overlook real symbol relationships, the constraints of context and location, the impact audience (market) has on decisions, and important qualitative/historical aspects of cartography. Such oversights are not necessarily inherent in computer use. But a special emphasis on turnkey systems produces a higher probability of overly simplistic views of map design and execution, and less ability to think creatively. This approach compels students to conform to a pre-existing, and deceptively inexact, body of knowledge. In turn, a graduate's on-the-job capabilities remain limited to a few 'user options' defined by system menus, which leads to less effective cartographic solutions. For the more academically inclined student, an emphasis on 'up-and-running' systems training induces a monolithic view of the subfield and a poor understanding of cartography's intellectually challenging potential.

Credentialism is rampant in academic cartography. What we have today is the professionalization of cartographic education and an increase in the student's desire to put mapping credentials on a résumé. Credentials exist to protect the public from incompetence, but often they merely conventionalize and regiment thinking so that education becomes training, while new talent and ideas are kept out of the practice.

It is ironic that the development of a cartographic professional elite, especially in the large government agencies established to oversee mapping, contributed to cartography's downfall as an integral part of geography, and encouraged the development of narrow, technical, and dull job-training courses (Muehrcke 1981). The agencies needed to know how to compile, design, and produce better maps, and university geography departments responded by becoming mini vocational schools for giving prospective employees the 'answers.'

Another pitfall of an applied cartography focus is that we run the risk of communicating the message that job-seeking students need only learn how to operate equipment to fulfill their goal. Understanding equipment operation, like all applied knowledge, is ephemeral, with little long-range utility. In addition, it seems unwise for instructors to tie students' education to the vagaries of the economy and the availability of jobs (Ford 1982). We have a responsibility to assist students in the development of knowledge they can use for much of their lives.

Finally, as expert systems develop, cartography students are beginning to see expertise as a function of large knowledge bases, knowledge decay rates, and legions of rules, but are less willing to progress beyond a machine's competency level. If this becomes commonplace – and I have seen signs of it among graduate students – reservoirs of true cartographic expertise and intelligence will begin to evaporate. That could place the future of cartography in jeopardy, instituting a descending spiral in the fund of cartographic knowledge. One way to retreat from the precipice may be to establish a 'national academic cartographic centre,' modelled after Europe's Cartography Department of the International Institute for Aerial Survey and Earth Sciences, where large numbers of students could receive training in applied cartography (Monmonier 1982, p. 100). To train small bands of credential-seeking university students in an applied cartography undergoing incessant change and of limited long-range utility only reduces potential cartographic knowledge acquisition.

Applied Cartography and 'The Industry'

Cartographic research and higher education have been guided to some extent by both private-sector and government needs, which has resulted in perils that seldom receive attention. First, there are risks involved in allowing 'the industry' to validate academic cartography, which is in effect what happens when government mapping agencies and private map companies, whether or not they handle government contracts, are permitted to become the sole 'market' for cartographic research (McNally 1987, p. 391; Petchenik 1985). The issues revolve around whether pragmatism should be the measure of all cartographic research. An excessively enthusiastic and, at times, arrogant market orientation reduces the scope of cartographic inquiry, and eliminates theoretical research that lacks immediate application.

Another problem is that technological change, and especially its rapidly increasing rate of acceleration, has had a marked impact on the applied-theoretical proportion in research because so few academic cartographers exist. Technology (and perceptions of the technical needs of map users) periodically threatens to overwhelm academic cartographers. Our typical response is to substitute long-term research goals for short-term monetary and technical gains. This puts American cartographers and their research in a position subordinate to national (i.e., military, corporate) interests. Taken to an extreme, academic cartographers could become a small corps of well-paid, automated mapping consultants – a technical farm system defined by the nonscholarly bureaucratic big leagues.

For three reasons, few insights from applied research find their way into cartography's core literature. Almost by definition, applied research attempts to provide solutions to short-term problems, not new interpretations or lasting knowledge. Such insights are better placed in technical government reports and conference proceedings. Also, the applied cartography community seldom avails itself of academic literature because it is seen as irrelevant or archaic. Finally, agency security restrictions placed on cartographic research (whether sensitive or not) further aggravates the access problem. It prevents academic access to most government activity, but security consciousness also restrains interagency contact, which causes massive duplications of effort and other inefficiencies.

The military-industrial complex remains a major influence in cartography, with the Department of Defense (DOD) being the main, but not only, point of contact. Academic cartographers have taken advantage of the military's recent 'spend-up,' especially in the area of software development, which is anticipated to comprise 10 percent of DOD's 1990 budget (Jacky 1985, p. 26). The U.S. Army research office offers special grants in cartography and remote sensing. In the 1950s, such grants (especially through the ONR) were designed partly to vitalize emerging cartography programs in geography departments in the eastern United States. However, current concerns are aimed more narrowly at the application of modern and futuristic technology to the problems of military mapping, navigation, and weapons control. To some degree, the continuing military influence on cartographic research, and the willingness of academic cartographers to become fascinated with missile guidance systems, elaborate tank and jet navigation devices, etc., parallels the military's unprecedented advance into many sectors of American society. Cartographers have studiously ignored morality issues in the subdiscipline.[3] For the time being, members of the ACSM should consider the code of ethics they adopted when joining (American Congress on Surveying and Mapping 1988, p. 43).

There remains a need for American academic cartographers to play their traditional 'watchdog' role. Because they are one of the few places where tradition is respected and revered, universities must endure as the bastions of tradition (Ford 1982). Theoretical support and evaluation of cartographic praxis still comes only from within the academy. Now, more than ever, it is important to retain this focus, and to pass it along to our students.

Cartography and Its Relationship to Geography

An overemphasis on applications-oriented cartographic research and training will aggravate certain extant disagreeable conditions in geography. One discipline-wide problem is the persistence of various dichotomies (James 1967). Geographers are not alone, of course, but oppositions that pit physical against human, regional against systematic, theoretical against applied, etc., are particularly debilitating in relatively small disciplines. Unfortunately, ours is actually a *tripartite* discipline, divided into camps of physical, human, and so-called methodologists or 'technical' geographers. Faculty and course specializations are classified accordingly in many geography departments, isolating students into one of the three niches. Cross-training in the three divisions is increasingly rare, especially among recent doctorates. Yet, an application-oriented view of cartography permits students trained in other geography subfields to label themselves cartographers, or GIS or AI specialists, as long as they have had some experience in mapping on a computer.[4] Along the way, they have acquired little sense of the literature or research traditions in cartography or allied fields.

Forthcoming volumes on the state of American geography will present the discipline in trichotomous terms. An AAG-sponsored project will consign cartography to a section on geographic methods (Abler, Marcus, and Olson Forthcoming), while a second text places cartography in a 'methodology' section with GIS, mathematical and statistical models, remote sensing, and microcomputer use (Gaile and Willmott 1989). Maps are the natural tool of geographers; their design, analysis, and comparison are part of a stock of research methods. But *cartography* is more than mere method. Such an oversight automatically purges the history of cartography at the very moment it may prove stimulating to the rest of the field (Harley 1988; Harley and Woodward 1987; Lewis 1987; Wood 1977). It locks out semiotic approaches to maps (Schlictmann 1985; Wood and Fels 1986), and an emergent concern with maps as mirrors of socio-cultural beliefs and values (e.g., Gilmartin 1984; Harley 1983, 1987; Nemeth 1987, 1988; Wood 1984; Woodward 1985), both of which are within the broader purview of human geographic research.

A trichotomous discipline perpetuates a view of cartography as solely a method or technique, which enhances the image of departmental cartographers as 'staff technicians.' Eagerness to see cartography as the quintessential applied subfield lends itself to asking a cartographer to bear the responsibility of automating departmental secretarial functions and/or bringing older faculty into the information age. Simultaneously, physical and human geography become labelled 'substantive' areas of research, while cartography is merely a nonsubstantive support segment of 'real' geography.

Moreover, friction between age cohorts in geography departments is aggravated: technical functions almost always are performed by application-oriented younger faculty; elder regionalists see their positions being usurped and the discipline disintegrating, even while they may avail themselves of the new technology (Hausladen and Wyckoff 1985). The situation in most geography departments today allows both groups to condescend toward the other.

Cartographers are as much at fault as anyone for the prevailing view of their research and departmental functions. They are the most aggressive to explore the possibilities that independence from geography might offer. And, indeed, if they conceive of their field as purely applied methods, then cartographers would likely feel more comfortable intellectually in computer science or engineering programs. But only if the applied, inherently technical, side of cartography is permitted to constitute the core can cartography be conceived as tangental to geography.

Concluding Comments

Despite shortcomings in the experimental paradigm, I do not advocate complete abandonment of theoretical research in cartography; that alternative leads to intellectual bankruptcy. Perhaps we should consider new conceptualizations of cartographic research in light of those proposed in other geography subfields.

Couclelis (1986) suggests a research framework for behavioral geography with interesting implications for cartography. In the recent past, behavioral geographers were attempting to produce high-level theoretical explanations. Inconclusive results caused abandonment of the pursuit in favor of applied research. An information processing paradigm has now supplanted earlier models, but Couclelis recognizes the philosophical and operational problems inherent in the machine-mind analogy. All of this should sound very familiar to cartographers. She suggests an alternative approach, whereby geographers produce 'meta-theories' to elaborate the *implications* of ways of defining, representing, and analyzing phenomena instead of dealing with the phenomena themselves. By developing a 'frame' to organize human experience, not to explain observed facts, such an approach may help define a more formal research agenda, which succeeding research could address at a higher level of theory formulation.

We are faced with the proliferation of applied training courses in cartography, and even separate specialties and degrees in applied cartography, at both under-graduate and graduate levels. These structures do not *educate* our students; they place cartographic instruction at the mercy of commercial and governmental interests. The obvious question of professional ethics has yet to be addressed: 'Is it proper for higher education priorities to be dictated by those with commercial interests at stake?' The applied trend appears to me to be an educational misfire, debilitating to further intellectual development of both cartography and geography. In short, we should not rearrange courses to suit ephemeral trends; cartography courses should be part of core curricula without alienating nontechnical students. Indeed one conception of geography locates cartography at the intellectual heart of the discipline, with other subfields interacting in symbiosis (King 1982).

By definition, cartographers must possess a strong applied approach to their field. But current priorities are weighted too heavily toward development and training in applications of short-term value, almost to the exclusion of research and education as a long-term investment. Cartography has adopted such an orientation before, as has the discipline of which it is a part. Perhaps applied cartography trends catalyze applied geography movements as a whole. If so, this 'indicator' subfield deserves increased scrutiny, especially with regard for its current perils.

Notes

1. It is worth noting that 'applied cartography' is mentioned consistently by European cartographers, but less so on the other side of the Atlantic. Perhaps this is because North American cartographers have long assumed cartography to be an applied

discipline, whereas even application-oriented European practitioners have explored the theoretical aspects of maps. Terminology serves to segment broad research and teaching responsibilities into constituent parts, but in the United States cartography has always been seen as inherently, and almost wholly, applied.

2. Cartographers have considered the human mind as analogous to an image-manipulating machine, with distinct units working away first at recognition, then discrimination, estimation, and comprehension tasks. If one thinks of the mind as a kind of pattern-deciphering machine, then it is easy to reverse the metaphorical elements and consider machines as carrying the potential to 'think.' This is precisely what fifth-generation computers are supposed to do, and cartographers are eager to discover if the human element can be removed from cartography via wholesale mechanization of spatial thinking. Cartographers are resourceful for adopting this metaphor from physical science (Minsky 1986), and a flurry of acronyms, specialized jargon, and manufactured terms (e.g., 'brainware' and 'geomatics') have been adopted in an effort to develop new concepts. But excitement gives way to blind zeal (Dahlberg and Jensen 1986; Taylor 1985). For example, although its prospects, meaning, and even its existence are open questions, artificial intelligence (AI) causes heart palpitations for some (Blades and Spencer 1986; Smith 1984).

3. Avoidance of moral issues may be ending in cartography. A long-awaited panel on 'Ethics in Cartography' was scheduled for the March, 1989 meeting of the Association of American Geographers in Baltimore, Maryland.

4. Cartography, remote sensing, and GIS are seen as nearly the same by many geographers and cartographers. A perusal of any recent copy of the AAG *Guide to Departments of Geography* reveals an increasing willingness to tag these research interests onto one's name. At a 1988 'GIS in University Education' conference at Ohio State University, the question of whether GIS should *assimilate* cartography, or vice-versa, was placed on the agenda despite lack of any theoretical base, or even an agreeable definition of GIS (Coulson 1988, p. 15).

References

Abler, R., Marcus, M., and Olson, J. Forthcoming. *Ties That Bind*.

American Congress on Surveying and Mapping. 1988. Code of ethics. ACSM *Bulletin* 114:43.

Association of American Geographers. 1949. Office of Naval Research (announcement). *Professional Geographer*, n.s. 1:27.

Association of American Geographers. 1953. Announcement of the Office of Naval Research contracts for research in geography. *Annals of the Association of American Geographers* 43:1–3.

Baker, O.E. 1921. The increasing importance of the physical conditions in determining the utilization of land for agricultural and forest production in the United States. *Annals of the Association of American Geographers* 11:17–46.

Baker, O.E. 1926. Agricultural regions of North America. *Economic Geography* 2:459–93.

Barnes, C.P. 1929. Land resource inventory in Michigan. *Economic Geography* 5:22–35.

Beishlag, G. 1951. Aims and limits in teaching cartography. *Professional Geographer*, n.s. 3(4):6–8.

Blades, M. and Spencer, C. 1986. The implications of psychological theory and methodology for cognitive cartography. *Cartographica* 23:1–13.

Castner, H.W. 1983. Research questions and cartographic design. In *Graphic Communication and Design in Contemporary Cartography* (Volume 2 of Progress in contemporary cartography series), ed. D.R.F. Taylor, pp. 87–113. New York: Wiley.

Couclelis, H. 1986. A theoretical framework for alternative models of spatial decision and behavior. *Annals of the Association of American Geographers* 76:95–113.

Coulson, M. 1988. GIS in university education: Columbus, Ohio, April 30–May 1, 1988 (announcement). Canadian Cartographic Association. *Newsletter* 14(1):15.

Dahlberg, R.E. 1978. Toward conceptual models of education and training in applied cartography. *Proceedings*, Applied Geography Conference, State University of New York-Binghamton.

Dahlberg, R.E., and Jensen, J.R. 1986. Education for cartography and remote sensing in the service of an information society: the U.S. case. *American Cartographer* 13:51–71.

Degani, A. 1980. Methodological observations on the state of geocartographic analysis in the context of automated spatial information systems. In *Map Data Processing*, ed. H. Freeman and G.Pieroni, pp. 207–221. New York: Academic Press.

Donnelly, D.P., ed. 1985. *The Computer Culture: A Symposium to Explore the Computer's Impact on Society*. Rutherford, NJ: Fairleigh Dickinson University Press.

Dreyfus, H.L., and Dreyfus, S.E. 1986. *Mind over Machine: The Power of Human Intuition and Expertise in the Era of the Computer*. New York: The Free Press.

Dunbar, G. 1981. Credentialism and careerism in American geography, 1890–1915. In *The Origins of Academic Geography in the United States*, ed. B. Blouet, pp. 71–88. Hamden, CN: Archon Books.

Eckert, M. [1908] 1977. On the nature of maps and map logic. trans. W. Joerg. In *The Nature of Cartographic Communication* (*Cartographica* Monograph No. 19), ed. L. Guelke, pp. 1–7.

Fenneman, N.M. 1931. *Physiography of Western United States*. New York: McGraw-Hill.

Finch, V.C. 1933. *Montfort: A Study in Landscape Types in Southwestern Wisconsin*. (Geographical Society of Chicago Bulletin No. 9.) Chicago: Geographical Society of Chicago.

Ford, L.R.1982. Beware of new geographies. *Professional Geographer* 34:131–135.

Gaile, G.L., and Willmott, C. 1989. *Geography in America*. Merrill: Columbus, OH.

Gilmartin, P. 1984. The austral continent on sixteenth century maps/ an iconological interpretation. *Cartographica* 21:38–52.

Goode, J.P. 1929. The polar equal area: a new projection for the world map. *Annals of the Association of American Geographers* 19:157–161.

Hanson, F. 1952. The applied cartographic program of the Army Map Service. *Professional Geographer*, n.s. 4:14.

Harley, J.B. 1983. Meaning and ambiguity in Tudor cartography. In *English Map-Making 1500–1650*, ed. S. Tyacke, pp. 22–45. London: British Library.

Harley, J.B. 1987. Innovation, social context and the history of cartography/ review article *Cartographica* 24:59–68.

Harley, J.B. 1988. Silences and secrecy: the hidden agenda of cartography in early modern Europe. *Imago Mundi*. 40:57–76.

Harley, J.B., and Woodward, D., eds. 1987. Cartography in prehistoric, ancient, and medieval Europe and the Mediterranean. In *The History of Cartography* (Vol. One). Chicago and London: University of Chicago Press.

Harrison, R.E. 1950. Cartography in art and advertising. *Professional Geographer*, n.s. 2(6):12–15.

Hausladen, G., and Wyckoff, W. 1985. Our discipline's demographic futures: retirements, vacancies, and appointment priorities. *Professional Geographer* 37:339–343.

Hudson, G.D. 1936. The unit area method of land classification. *Annals of the Association of American Geographers* 26:99–112.

Imhof, E. [1963] 1977. Tasks and methods of theoretical cartography. trans. E. Spiess and M. Merriam. In *The Nature of Cartographic Communication* (*Cartographica* Monograph No. 19), ed. L. Guelke, pp. 26–36.

Jacky, J. 1985. The 'Star Wars' defense won't compute. *The Atlantic* 225(6):18–30.

James, P.E. 1967. On the origins and persistence of error in geography. *Annals of the Association of American Geographers* 57:1–24.

Jenks, G.F. 1953. An improved curriculum for cartographic training at the college and university level. *Annals of the Association of American Geographers* 43:317–331.

Jones, W.D. 1930. Ratios and isopleth maps in regional investigation of agricultural land occupance. *Annals of the Association of American Geographers* 20:177–195.

Jones, W.D., and Finch, V.C. 1925. Detailed field mapping in the study of the economic geography of an agricultural area. *Annals of the Association of American Geographers* 15:148–157.

Keates, J.S. 1982. *Understanding Maps*. New York: Wiley.

King, R. 1982. On geography, cartography and the 'fourth language'. *Geographical Research Forum* 5:42–56.

Lewis, G.M. 1987. Misinterpretation of Amerindian information as a source of error on Euro-American maps. *Annals of the Association of American Geographers* 77:542–563.

Lobeck, A.K. 1924. *Block Diagrams and Other Graphic Methods Used in Geology and Geography*. New York: Wiley.

Lobeck, A.K. 1933. *Airways of America*. New York: Geographical Press.

Mackay, J.R. 1954. Geographic cartography. *Canadian Geographer* 4:1–14.

McNally, A. 1987. 'You can't get there from here' with today's approach to geography. *Professional Geographer* 39:389–392.

Martin, L. 1930. The Michigan-Wisconsin boundary case in the Supreme Court of the United States. *Annals of the Association of American Geographers* 20:105–163.

Miller, O.M. 1931. Planetabling from the air: an approximate method of plotting from oblique aerial photography. *Geographical Review* 21:201–212; 660–662.

Minsky, M. 1986. *The Society of Mind*. New York: Simon and Schuster.

Monmonier, M.S. 1982. Cartography, geographic information, and public policy. *Journal of Geography in Higher Education* 6:99–107.

Monmonier, M.S. 1985. *Technological Transition in Cartography*. Madison, WI: University of Wisconsin Press.

Muehrcke, P.C. 1981. Whatever happened to geographic cartography? *Professional Geographer* 33:397–405.

Nemeth, D.J. 1987. The architecture of ideology: Neo-Confucian imprinting on Cheju Island, Korea. *University of California Publications in Geography* 26.

Nemeth, D.J. 1988. Topomancy – The first tradition of geography? Paper presented at the annual meeting, Association of American Geographers, Phoenix, AZ, April 6–10.

Odell, C.B. 1950. Cartography and cartographers in commercial map companies. *Professional Geographer*, n.s. 2(6):17–19.

Petchenik, B. 1983. A map maker's perspective on map design research 1950–1980. In *Graphic Communication and Design in Contemporary Cartography* (Volume 2 of Progress in contemporary cartography series), ed. D.R.F. Taylor, pp. 37–68. New York: Wiley.

Petchenik, B. 1985. Maps, markets, and money: a look at the economic underpinnings of cartography. *Cartographica* 22:7–19.

Philbrick, A.K. 1953. Toward a unity of cartographical forms and geographical content. *Professional Geographer*, n.s. 5:11–15.

Quam, L.O. 1946. Cartography for geographers. *Professional Geographer* (old series) 4:10–12.

Rabenhorst, T.D., and McDermott, P.D. 1988. *Applied Cartography: Maps and Imagery as Source Data*. Merrill: Columbus, OH.

Raisz, E. 1950a. Introduction. *Professional Geographer*, n.s. 2(6):9–11.

Raisz, E. 1950b. Report of the committee on cartography. *Professional Geographer*, n.s. 2(2):15–16.

Raisz, E. 1952. Cartography. *Professional Geographer*, n.s. 4(2):12.

Ratajski, L. 1973. The research structure of theoretical cartography. *International Yearbook of Cartography* 13:217–227.

Ridge, T.L. 1948. Naval interests in geographical research. *Professional Geographer* (old series) 7:17–20.

Ristow, W.W. 1983. Cartography and Robinson then and now. SLA, Geography and Map Division *Bulletin* 132:8–16.

Robinson, A.H. 1951. University training for government cartographers. *Professional Geographer*, n.s. 3(4):4–6.

Robinson, A.H. 1952. *The Look of Maps: An Examination of Cartographic Design*. Madison, WI: University of Wisconsin Press.

Robinson, A.H. 1954. Geographic cartography. In *American Geography: Inventory and Prospect*, ed. P.E. James and C.F. Jones, pp. 553–577. Syracuse: Syracuse University Press.

Robinson, A.H., and Petchenik, B.B. 1976. *The Nature of Maps: Essays Toward Understanding Maps and Mapping*. Chicago and London: University of Chicago Press.

Roszak, T. 1986. *The Cult of Information: The Folklore of Computers and the True Art of Thinking*. New York: Pantheon Books.

Sauer, C.O. 1919. Mapping the utilization of the land. *Geographical Review* 8:47–54.

Schlictmann, H. 1985. Characteristic traits of the semiotic system 'map symbolism'. *Cartographic Journal* 22:23–30.

Smith, G-H. 1928. The settlement and distribution of population in Wisconsin: a geographical interpretation. *Geographical Review* 18:402–421.

Smith, T.R. 1984. Artificial intelligence and its applicability to geographical problem solving. *Professional Geographer* 36:147–158.

Taylor, D.R.F. 1985. The educational challenges of a new cartography. *Cartographica* 22:19–37.

Voskuil, R.J. 1950. Cartographers in government. *Professional Geographer*, n.s. 2:29–32.

Wellar, B.S. 1985. The significance of automated cartography to an information society. *Cartographica* 22:38–50.

Wood, D. 1977. Now and then: comparisons of ordinary Americans' symbol conventions with those of past cartographers. *Prologue*, The Journal of the National Archives 9:151–161.

Wood, D. 1984. Cultured symbols/ thoughts on the cultural context of cartographic symbols. *Cartographica* 21:9–37.

Wood, D., and Fels, J. 1986. Designs on signs/ myth and meaning in maps. *Cartographica* 23:54–103.

Woodward, D. 1974. The study of the history of cartography: a suggested framework. *American Cartographer* 1:101–115.

Woodward, D. 1985. Reality, symbolism, time, and space in medieval world maps. *Annals of the Association of American Geographers* 75:510–521.

Wright, J.K. 1936. A method of mapping densities of population: with Cape Cod as an example. *Geographical Review* 26:103–110.

Robert A. Rundstrom
Public Affairs/Geography
George Mason University
Fairfax, VA 22030
U.S.A.

13. Women in Applied Geography

A long time ago, in my first college geography class, I was told 'geography is what geographers do,' and this casual remark stuck with me. Thus I have never understood the need for the term 'applied geography.' Geography is what geographers do, whether they do it in the classroom for the love of learning, or in government agencies for the good of the country, or in private enterprise for profit. The title 'applied geographer' seems unnecessary; professional geographer or practitioner is more appropriate. Members of other disciplines, wherever they are employed, seem to refer to themselves simply as sociologists, anthropologists, or economists.

The term applied geography is here to stay, however, and the purpose of this chapter is to examine some of the career opportunities and satisfactions that women have found in it and the contributions they have made. After a brief background section, the larger part of the chapter is historical, based on selected case studies. It describes the successful careers of five women who were active in applied geography before the days of affirmative action and who accomplished some remarkable things. The concluding section will briefly look at women in geography today.

Background

Today the whole discipline seems to be trying to clarify its feelings about the applied sector. There is regrettably a tendency for some geographers in academia, devotees of 'pure research,' to look upon applied geography as somehow a lesser branch of the discipline. Certainly the women in applied geography are well aware of this attitude, as revealed in numerous comments, letters, and survey answers. Such an attitude is reminiscent of that held by members of the British aristocracy toward people 'in trade,' and is not diminished by the fact that geographers in business and government often earn more money than professors.

Disciplines flourish according to the degree of respect accorded them by the general public and according to the perceived need for applications of those disciplines. Those that are 'in' at any time, the favorites of deans and recipients of many grants, are those whose applications are most immediately obvious, publicized, and glamorous. Unfortunately, our own discipline is usually not 'in.' We are striving to improve the situation by remedying some of the appalling deficiencies in geographic education. It would help if the applications of geography were better known and its practitioners more clearly identified.

Geographers are handicapped by the lack of a clear and lasting identity, particularly in the applied sector; in the academic world we are reasonably safe. (I say reasonably, because I teach geography in a Department of Public Affairs, where my

M. S. Kenzer (ed.), Applied Geography: Issues, Questions, and Concerns, 193–204.
© 1989 *Kluwer Academic Publishers.*

colleagues and I are all professors of Geographic and Cartographic Sciences; this sounds more high-tech than plain old Geography.) In applying their discipline to real world problems, geographers are often renamed. Geography gets lost in urban planning or environmental science out in the workaday world, just as it gets lost in social science and earth science in grade school. An economist working on a development plan is still an economist; a geographer is apt to be rechristened planner, analyst, or cartographer. It is unfortunate for the discipline that people applying geography have to assume other identities. It is one of the problems that women in applied geography mention most frequently.

Although the public often fails to recognize their work as applied geography, geographers are, by virtue of their training, effective coordinators and team leaders for multidisciplinary work in business and government. Many women have commented on the fact that their education in geography prepared them to think integratively and holistically; they see the forest as well as the trees.

Women geographers have long been forced to think creatively when it comes to finding employment. Before the days of affirmative action, it was difficult for them to find academic jobs. Many women still feel that getting tenure is more difficult for them. Women will always be well represented in the applied sector. Like men, they tend to be practical, they are good at time management, they are accustomed to juggling many demands at once, and they want to make money.

Typically, women studying geography also think of adding another credential, another string to their bow. Today it may well be computer science. Certainly they tend to emphasize the more technical subfields within geography – e.g., remote sensing, cartography, and statistical methods. But, it may also be a foreign language. For example, at George Mason University, I have had several women majors who very consciously double-majored in geography and Russian, knowing that they had thus enhanced their employment possibilities with the federal government. Another woman major found immediate and well-paid employment because of her combined knowledge of remote sensing and French. This is carrying on a long tradition for women in applied geography. In the pre- and post-World War II era, women taking geography degrees were urged to add another credential, something practical, something that could be used anywhere (which typically meant wherever their husbands' jobs carried them). These urgings came from practical mothers, other women, and (yes) many male geography professors. The useful credentials were usually in secondary education or library science, as is illustrated in the case studies below. The pattern continues, though library science has been replaced by computer science and education by business administration.

Five Women

When geographers began to apply their disciplinary skills to practical problems, women were among the first to do so. Many people place the beginnings of applied geography in the U.S. in the 1930s with the work of G. Donald Hudson, Malcolm Proudfoot and others for the Tennessee Valley Authority (James and Martin 1981).

It may be surprising to hear about the career of Dr. Helen Strong, who was already working in Washington during World War I. By 1930 she was well enough known to be the subject of a full spread in the *Washington Post*, a story entitled 'Linking Geography with Business' that described her work as chief geographer of the United States Bureau of Foreign and Domestic Commerce (Washington Post 1930). Probably only specialists in the history of cartography, or map librarians, are aware of the great contributions of Clara E. LeGear to these fields. Her long career in the Map Division at the Library of Congress began in 1914. It is true, however, that the participation of substantial numbers of women in applied geography began during World War II, and it was in this period that the careers of Evelyn Pruitt and Betty Didcoct Burrill began. Shortly afterward, in the early 1950s, Dorothy A. Muncy received her Ph.D. from Harvard and began a career as an urban planner and industrial land use consultant.

The Washington metropolitan area, with its climate of flourishing applied geography, has long been an employment Mecca for women geographers. These five women spent most of their careers there. That was not always the way they had planned it, however.

Helen Strong

Helen Strong (1890–1973) was the first woman to receive a Ph.D. in geography from the University of Chicago (1921) and was the first woman on the United States Geographic Board (later called the U.S. Board on Geographic Names), a position to which she was appointed by President Coolidge. She successively and successfully held posts in the Bureau of Foreign and Domestic Commerce (in the Department of Commerce), Soil Conservation Service (SCS), (in the Department of Agriculture) and Office of the U.S. Army Chief of Staff. She was credited with linking geography with business and helping to make big business bigger. She predicted in 1930 that the businessman of the future would have a geographic background facilitating scientific salesmanship in all parts of the world. Alas, surveying the scene in the 1980s, we have to give her poor marks as a prophet. She was a good geographer and a good advocate of geography, though, and all her 'firsts' helped pave the way for the employment of many other geographers in the federal government.

Her first involvement with Washington came during World War I. Having studied at Chicago under J. Paul Goode, Rollin Salisbury, Harlan H. Barrows, and Ellen Churchill Semple, she was finishing up her master's degree when she was asked to come to Washington to work for O.E. Baker in the Department of Agriculture. Baker was cooperating with Isaiah Bowman, then-Director of the American Geographical Society. Bowman was in charge of geographic research dealing with territories and boundaries in preparation for the peace conference that would take place when the war ended. The young Helen learned from Baker to use and analyze all sorts of domestic and foreign statistics; she also made contacts with people who later came to be influential in the SCS, and the U.S. Geological Survey.

Out of these experiences came her ambition to do for manufacturing statistics in the U.S. what O.E. Baker was doing with agricultural statistics; she was convinced that geography had much to contribute to business and industry.

After the war she returned to Chicago and completed a doctorate in geography, writing a dissertation on the manufacturing geography of Cleveland that certainly did not foreshadow her later heavy involvement with exotic placenames (Strong 1958). She took a faculty position at the University of Missouri (1921–1923), but on a chance visit to Washington received an offer to become Geographer of the Bureau of Foreign and Domestic Commerce. At this point Helen ran into a problem that has been with geographers ever since: no one knew which Civil Service examination to give her. No specific exam existed for a professionally trained geographer to serve commerce, industry, and business.

Dr. Strong's work was primarily research and writing for the publications of the Bureau on foreign trade. She gradually saw to it that a geographic slant was incorporated into all of them. She made the map publishing industry her special point of contact and initiated the publication of *Geographic News*, a monthly publication that summarized place name changes, new map publications, and other information that came to her from the network of commercial attachés abroad. This led to her appointment to the United States Geographic Board, which was charged with the systematization of placename spellings.

Throughout her long career, Helen Strong continued to add to her list of firsts – some firsts for a woman, some for a geographer, usually both. One speaking engagement that remained in her memory was before the dignified Export-Import Club of Philadelphia, their first experience with a woman speaker and a geographer. An example of her many significant contributions is her action in persuading Secretary of Commerce Herbert Hoover to order a switch from Mercator to Goode's Homolosine maps for Department of Commerce publications. Analyzing statistics was not her only work. She was once consulted by the export branch of the Jantzen swimsuit company concerning seasons and conditions on bathing beaches around the world, information that she readily supplied.

Geographic work at the Bureau ended with a two-thirds reduction in force in 1933, at the depth of the Depression. Helen Strong accepted a position in the U.S. Coast and Geodetic Survey, where she learned a lot about charts. But soon, at the behest of Dr. Hugh S. Bennett, Director of the recently organized Soil Conservation Service, she moved again. Her task was to bring the SCS into cooperation with schools, colleges, and universities throughout the country, ensuring that students at all levels were exposed to the importance of soil conservation. During the next few years, her fieldwork took her into forty-six states, and she enlisted the aid of her colleagues in geography departments throughout the country.

She also acquired a lot of knowledge from the engineering sections of the SCS, knowledge that proved useful in her last position in applied geography in the Office of the Chief of Staff, U.S. Army. World War II had produced an acute need for physical geographers who could apply their expertise in climatology, terrain analysis, and hydrology to strategic problems. Helen Strong turned her many talents, which by now included air photo interpretation, to the writing of studies in military geography.

She lived for many more years after her retirement and continued to be professionally active, teaching part time at Elmhurst College. She was a great promotor of geography and of women in geography. Although her published work consisted largely of government reports, it included articles in *Economic Geography* and the *Annals* (Strong 1936, 1937).

Clara LeGear

Clara Egli LeGear's (b. May 2, 1896) career in applied geography in Washington began even earlier than Helen Strong's and lasted even longer. It was a very different kind of career, however, and her contributions were very different. While Helen was really the 'compleat geographer,' moving from agency to agency and from challenge to challenge, demonstrating that geography had a role everywhere, Clara was a specialist, steadily working away for fifty-eight years in the Library of Congress to produce her great contribution to historical cartography, Volumes 5 through 9 of the monumental *List of Geographical Atlases in the Library of Congress* (LeGear 1958, 1963, 1973, 1974, 1988). She was a full-time employee of the Library for forty-seven years, retiring in 1961, only to be appointed Honorary Consultant in Historical Cartography, a position she held for eleven more years.

Clara Egli came to the Library in 1914 to work in the Cataloging Division, but she transferred to the Map Division only eleven months later and was associated with it for the rest of her long career. Hired as a typist – the story of so many women of her generation – she moved through almost all phases of map librarianship: cataloging, reference, acquisitions, bibliography, and administration. She continued her education, acquiring her bachelor's (1930) and master's (1936) degrees in Library Science from George Washington University and taking additional courses in cartography and editing at the U.S. Department of Agriculture Graduate School.

By 1945, Clara Egli LeGear had been with the Library over thirty years and had been Assistant Chief of the Geography and Map Division for fourteen. Her most productive years were about to begin, however, as she moved into full-time writing and bibliographic activities. Her first major publication came in 1949: *Maps: Their Care, Repair and Preservation in Libraries*, which has become a classic in map librarianship (LeGear 1949). She was named Bibliographer in 1950 and resumed work on *A List of Geographical Atlases in the Library of Congress*, which had been begun by Philip Lee Phillips, the first head of the Map Division. Clara had worked for Phillips from 1914 until his death in 1924. He had completed four volumes. She completed Volumes 5–8 during the remainder of her career as a regular Library employee and the eleven years as Honorary Consultant in Historical Cartography. Volume 9, the Comprehensive Author List for the eight volumes published from 1909 to 1974, was also compiled by Mrs. LeGear and is currently in press (LeGear In press).

Clara LeGear's contributions to historical cartography earned her the title, 'patron saint of maps.' Through her bibliographic efforts she shared her immense knowledge of the map collections of the Library of Congress with scholars the world over. She also was active in building up those collections. She was elected to the Association of American Geographers (AAG) in 1943 and received the Meritorious Achievement Award of the AAG in 1952. She received the Honors Award of the Special Library Association's Geography and Map Division (1957), the American Library Association's C.S. Hammond Award (1963), and the Library of Congress Distinguished Service Award (1963). In 1968 she was designated an Honorary Fellow of the American Geographical Society of New York.

Evelyn Pruitt

Evelyn Pruitt (b. April 25, 1918) must surely be one of the most honored women and one of the best known names in applied geography. In 1984 she received the James R. Anderson Medal of Honor from the Applied Geography Specialty Group of the AAG, the highest recognition in the field. This award was added to the honors already bestowed on her by Louisiana State University (1983), the American Society of Photogrammetry (1976), the Society of Woman Geographers (1981), and the Association of American Geographers (1972). She has been editor of the *Professional Geographer* (1957–1960), president of the Coastal Society (1976–1978), and a Regents Professor at UCLA. Evelyn, like Helen Strong, accumulated a lot of 'firsts' in the course of her long and distinguished career. She was, for example, both the first woman and the first geographer to be honored by LSU with the honorary degree, Doctor of Humane Letters (1983).

Evelyn's career is so recent, so well documented, and so well known to geographers that there is no need for this chapter to repeat the details (AAG 1984). I want to stress two things: first, the fact that she was involved in applied geography relatively early, in the wave of geographers who went to Washington in World War II; and second, the lasting nature of her contributions to applied geography. To a whole generation of geographers, Evelyn became Ms. Applied Geography.

A native Californian, Evelyn was born in San Francisco and received both her bachelor's (1940) and master's (1943) degrees in geography at UCLA. She came to Washington in 1942 as a cartographic editor with the Coast and Geodetic Survey, but by 1948 had moved to the Office of Naval Research (ONR) where she continued a productive career. She initiated and organized the first ONR program in geography and was the first to analyze Navy requirements for geographic knowledge. She essentially created the field of coastal geography, defined its objectives, and focused attention on the critical environments of world coasts (AAG 1984). She actively participated in the organization of the Coastal Studies Institute at Louisiana State University and chaired the Gulf Coast Committee of the Army Corps of Engineers' Shoreline Erosion Advisory Panel.

She is probably best known, however, as a pioneer in the field of remote sensing of the environment. Evelyn is actually credited with coining the term 'remote sensing,' thus naming one of the most important subfields of applied geography

(Pruitt 1979).

In 1971, Evelyn had an interesting observation about the status of women: 'The status of women in the Navy Department was illuminated when it was decided to bring together the "executives" (people with supergrades) to meet the new Assistant Secretary for Manpower. Only two women qualified, both from the Office of Naval Research – the newly appointed ONR Comptroller, and me. Perhaps there is a source of professional pride in that a geographer is the top-ranking woman scientist in the Navy.'

Betty Didcoct Burrill

It would certainly be an omission not to include a representative of the women who pioneered in the new field of geographic intelligence during and after World War II, although most of their work and virtually all their written output remains classified. When Betty Didcoct (b. September 20, 1915) retired from the Central Intelligence Agency (CIA) in 1972, after a government career of twenty-eight years, she was named as one of the first recipients of the new Career Intelligence Medal.

Betty Didcoct grew up in Nashville, with the modest ambition of becoming a famous archaeologist. Her undergraduate degree (1938) at Mount Holyoke College, with majors in archaeology and geology, was followed by two degrees from George Peabody College, a degree in library science (1939) and an M.A. in geography (1944) under J. Russell Whittaker. By this time she had decided that she wanted to teach college geography, where there was a real need resulting from the depletion of faculties by the war. Then S. Whittemore Boggs, The Geographer at the State Department, came to Peabody to visit the Geography Department. He needed someone with Betty's background in geography and library science and asked her to come to Washington. She figured that some government experience was bound to be useful in an academic career in geography, and so she slid into a different career path.

At the State Department she was given the initial task of organizing a large but motley collection of geographic materials so that the division's ninety researchers, including both geographers and cartographers, could use them effectively. It startles geographers today to learn that the working tools of the World War II era often left much to be desired. Betty once worked late into the night, completing an atlas of tinted hectograph maps that had been made for President Roosevelt to take to Yalta the next day.

After the war many geographers from the Office of Strategic Services (OSS) came to the State Department, and the geographic work was reorganized regionally. Betty elected to go with the Latin America group; she liked the region and had fieldwork experience there from her archaeology days. During the period of postwar readjustment, the CIA was created, and a number of geographers, including Betty, moved over to the new agency. By the early 1950s she had become Chief of the Western Hemisphere branch and was there for the rest of her career. Her group did basic geographic research on the whole hemisphere, from pole to pole. In practice, however, her research was focused on Latin America – an area badly in need of geographic analysis – with occasional attention to problem areas in other parts of the hemisphere.

In 1949 she was a member of the U.S. delegation to the first Consultation of the Geography Commission of the Pan-American Institute of Geography and History in Rio de Janeiro. A highlight of that trip was a fieldtrip with soils expert Robert Pendleton and conservationist William Vogt, to observe the techniques of large-scale soil reclamation that were being pioneered on worn-out coffee *fazendas*. Another Latin American conference that she remembers as a highlight was the Regional Conference of the International Geographical Union, meeting concurrently with a UNESCO Symposium on the Geoecology of Mountainous Regions of Tropical America, held in Mexico City in 1966.

She became active in research on the Antarctic during the International Geophysical Year (1957–58) and later was a member of the U.S. delegation to the Antarctic Treaty meeting in Washington (1959). She is pleased about her participation in that historic treaty – which enabled Antarctic research to go on without political overtones – and believes that it may be a model for future treaties on research in outer space.

Betty named another contribution to applied geography that she thinks has been a real success story. She was the person who first conceived the idea for JIG, the 'Jobs in Geography' section of the *AAG Newsletter*. Betty was active on the Career Placement Committee at a time when its work was largely matching up letters written in by geographic job-seekers with those sent in by prospective employers. The whole field of applied geography was opening up, Betty recalls, with more and more jobs in marketing, city planning, and industrial location, as well as in government and academia. She proposed that the laborious 'match-making' role of the Committee should be replaced with published advertisements by both employee and employer, and so JIG was born.

Dorothy A. Muncy

The fifth woman among these female pioneers in applied geography characterizes herself as a city planner and industrial geographer. Her career has been as a private consultant and her degrees are all post-World War II. Like the others, however, her career ambitions and goals were defined during a brief stint as a government employee; and, like the others, she wound up living in Washington. Working out of Washington, she has done consulting all over the country.

Dorothy A. Muncy (b. March 27, 1917) grew up in Atlantic City and still remembers the feel of living beside an ocean and not being surrounded. That feeling for open space carried over into her later planning work. She gained a strong interest in the social sciences in two years of undergraduate work completed at the University of Chicago before World War II. Right after high school she had worked in a public housing office run by the Works Progress Administration (WPA), perhaps one of the best remembered of the 'alphabet agencies' of the New

Deal. After Chicago she had an opportunity for a job with the WPA in its Washington headquarters, followed by a stint in its regional office in Chicago. She then went back to Washington during the war to work with the War Production Board, studying the sudden need for women workers, and devising ways to help keep women and older workers on the job and productive. Then she followed her husband, an electronics engineer, when he was transferred to Cambridge, Massachusetts, to work on radar development.

Many things from her past education and work experience must have come together in her mind at about this time, resulting in a strong desire to make a career out of planning spaces for people to live and work under optimum conditions. She describes herself as wandering about Harvard, investigating its educational riches, knocking on doors to ask about city planning. She spent the ensuing years acquiring her bachelor's (1947), master's (1949) and Ph.D. (1953) degrees in city planning, with a strong geography component. For the Ph.D. she was required to major in an arts and sciences discipline, which was geography.

She worked with Edward L. Ullman, Derwent S. Whittlesey, and Edward A. Ackerman, but was most influenced by Ullman. She thought that he asked all the right questions, questions that helped her put her interests into perspective. It was Ullman who pointed out that the boardwalk back in Atlantic City was a facade, a facade that protected the economic base of the resort town; planners must always pay practical attention to the things that bring in money. From Whittlesey she learned 'how to see,' how to read the landscape, and from Joseph A. Russell, a visiting professor one summer, she got a firm geographic basis in manufacturing and transportation. Gradually her major research interest evolved: a conviction that industry was losing good sites because its land needs were not known and explicit (Muncy 1954a).

Muncy's doctoral dissertation, *Industrial Land and Space Requirements*, was later published in a shorter version by the Urban Land Institute (Muncy 1954b). Her first client was a joint planning commission in Montgomery/Prince George's County (in the Maryland suburbs of Washington) assessing prospects for industrial development. With that report she established her principles: always tell the client the clear truth, without compromise, and remember that there is no such thing as a bad industry, only one improperly located. Many of her later clients were local governments and port authorities. She has been particularly interested in needs for parking and loading, thinking ahead to what the trucks of the future will be like. Another interest is in land use of shorelines, particularly with reference to deep-water locations. One of her studies was of the northern Chesapeake, showing how very few deep water sites suitable for certain industries still remain. A major consultant job took her to Brazil to help in planning a new deep-water port.

Dorothy Muncy is semi-retired but still active. When I interviewed her for this chapter she interspersed her career history with advice to women in applied geography. At the heart of her advice is this: geographers have something that planners do not: their knowledge of resources. A woman in planning or geography who today knows something about water transport and who follows trends in shipbuilding is in a unique position to build a career for herself.

Conclusions

Having looked at the diverse careers of five women, all of whom found intensely satisfying careers in applied geography and made distinctive contributions to their fields, we come back to the present. How many female geographers are currently working in the applied field, what are they doing, and how do they feel about it?

The total AAG membership in 1986 was 5787; of these, 1171 members (23.5 percent) were female (AAG 1987). That percentage has not changed much over the past five years. Of more relevance to this chapter is the breakdown of the membership by sex and occupation. Women are very much underrepresented among those employed by colleges and universities (making up only 13.5 percent of this group). They make up 35 percent of the student membership, a considerable overrepresentation. Something happens to them between the student stage and the professor stage. They do not all go to work for the government; the percentage of women in federal and other government is about the same as in the total membership. They are slightly overrepresented in private industry (26.9 percent), but they are definitely overrepresented in research centers (29.2 percent), nonprofit organizations (32.2 percent) and among the self-employed (29.8 percent). The employment patterns for women geographers clearly differ from those of men. Looking at the figures another way, we see that of the total of 1171 female members of the AAG, 28 percent are employed by colleges and universities, 10 percent by various levels of government, and approximately 17 percent by the private sector (AAG 1987).

It seems safe to say that applied geography is an important option for women majoring in geography. What has changed since the days of the five women whose outstanding careers were detailed above is the relative importance of the private sector. No longer is the federal government the major employer of professional geographers. Women do not necessarily all wind up in Washington, either, though it still has perhaps the greatest array of job possibilities.

The job titles held by women in the applied sector are varied. Regrettably, they seldom include the word geographer. Of the ninety-nine women responding to a recent survey, only six actually listed geographer as a title (Andrews and Moy 1986). The words most frequently used in the job titles were 'planning,' 'cartographer,' and 'environmental,' in combination with other words such as analyst. Evidently the adaptability of geography is still stronger than its identity. The women responding to the survey, and women talking to each other, are still saying that it is wise to add another string to the bow; get technical skills and management skills.

What is it that geography has to offer to the private sector and to government at all levels? From the careers of the five women described and from women currently active, it seems clear that the specific knowledge and techniques that can be applied to problem solving vary with economic and political conditions and with advancing technology. Area expertise, or regional specialization, so important in the two world wars, went into a period of decline but may again be needed. Skills in the graphic display of data have advanced from manual cartography to the sophisticated geographic information systems and remote sensing systems of today. The core of geography has remained its strength, however. For women (and men) contemplating careers in applied geography, their strong points are their ability to analyze data that vary spatially and their ability to synthesize and integrate data from different disciplines. Whether or not they are called 'applied geographers,' women can find satisfying and remunerative careers in applying their geographic knowledge and skills to the solution of social, political, economic, and environmental problems. Like the five women described above, they will continue to stake out new claims for geography.

Acknowledgment

I wish to express appreciation to the Society of Woman Geographers, 1619 New Hampshire Avenue NW, Washington, DC, for allowing me to use and quote from its files. All unattributed quotations in this paper are from these files.

References

AAG. 1984. Pruitt Awarded James R. Anderson Medal of Honor. *AAG Newsletter*. 19 (June 1984):6.

AAG. 1987. Membership increases almost 5% from 1985. *AAG Newsletter* 22 (March 1987):9.

Andrews, Alice C., and Moy, Kate. 1986. Women geographers in business and government: a survey, *Professional Geographer* 38:406–410.

James, Preston E. and Martin, Geoffrey J. 1981. *All Possible Worlds*. Second Edition. New York: Wiley.

LeGear, Clara Egli. 1949. *Maps: Their Care, Repair, and Preservation in Libraries*. Washington, DC: Library of Congress.

LeGear, Clara Egli. 1958. *A List of Geographical Atlases in the Library of Congress*. Vol. 5: Titles 5325–7623. Washington, DC: Library of Congress.

LeGear, Clara Egli. 1963. *A List of Geographical Atlases in the Library of Congress*. Vol. 6: Titles 7624–10254. Washington, DC: Library of Congress.

LeGear, Clara Egli. 1973. *A List of Geographical Atlases in the Library of Congress*. Vol. 7: Titles 10255–18435. Washington, DC: Library of Congress.

LeGear, Clara Egli. 1974. *A List of Geographical Atlases in the Library of Congress*. Vol. 8: Index to Volume 7. Washington, DC: Library of Congress.

LeGear, Clara Egli. In press. *A List of Geographical Atlases in the Library of Congress*. Vol. 9: Comprehensive Author Index. Washington, DC: Library of Congress.

Library of Congress. 1973. 58 years of service to LC completed by Mrs. Clara LeGear. *Information Bulletin*. January 26, 1973, 29.

Library of Congress. 1974. Clara E. LeGear interviewed by Walter W. Ristow and John Cole, April 19, 1974. LC Tape Call No. LWO7798.

Muncy, Dorothy A. 1954a. Land for industry: a neglected problem. *Harvard Business Review* 32:51–63.

Muncy, Dorothy A. 1954b. *Space for Industry: Analyses of Site and Location Requirements*. Urban Land Institute Technical Bulletin 23. Washington: ULI.

Muncy, Dorothy A. 1970. Planning guidelines for industrial park development. *Urban land*. 29:3–10.

204   ALICE ANDREWS

Pruitt, Evelyn. 1979. The office of naval research and geography. *Annals, Association of American Geographers* 69:103–108.
Society of Woman Geographers. 1974. Betty Didcoct (Burrill) interviewed by Delia Goetz, April 4, 1974.
Strong, Helen M. 1936. Regionalism: its cultural significance. *Economic Geography*. 393–410.
Strong, Helen M. 1937. Regions of manufacturing intensity in the United States. *Annals, Association of American Geographers* 27:23–43.
Strong, Helen M. 1958. Adventures in geography. Unpublished address presented at the annual dinner, Chicago Group, Society of Woman Geographers, April 18, 1958.
Washington Post. 1939. Linking geography with business. Sunday, August 5, 1930:5.

Alice Andrews
Public Affairs/Geography
George Mason University
Fairfax, VA 22030
U.S.A.

Index

The GeoJournal Library

1. B. Currey and G. Hugo (eds.): *Famine as Geographical Phenomenon.* 1984
   ISBN 90–277–1762–1
2. S. H. U. Bowie, F.R.S. and I. Thornton (eds.): *Environmental Geochemistry and Health.* Report of the Royal Society's British National Committee for Problems of the Environment. 1985 ISBN 90–277–1879–2
3. L. A. Kosiński and K. M. Elahi (eds.): *Population Redistribution and Development in South Asia.* 1985 ISBN 90–277–1938–1
4. Y. Gradus (ed.): *Desert Development.* Man and Technology in Sparselands. 1985 ISBN 90–277–2043–6
5. F. J. Calzonetti and B. D. Solomon (eds.): *Geographical Dimensions of Energy.* 1985 ISBN 90–277–2061–4
6. J. Lundqvist, U. Lohm and M. Falkenmark (eds.): *Strategies for River Basin Management.* Environmental Integration of Land and Water in River Basin. 1985 ISBN 90–277–2111–4
7. A. Rogers and F. J. Willekens (eds.): *Migration and Settlement.* A Multi-regional Comparative Study. 1986 ISBN 90–277–2119–X
8. R. Laulajainen: *Spatial Strategies in Retailing.* 1987 ISBN 90–277–2595–0
9. T. H. Lee, H. R. Linden, D. A. Dreyfus and T. Vasko (eds.): *The Methane Age.* 1988 ISBN 90–277–2745–7
10. H. J. Walker (ed.): *Artificial Structures and Shorelines.* 1988
    ISBN 90–277–2746–5
11. A. Kellerman: *Time, Space, and Society.* Geographical Societal Perspectives. 1989 ISBN 0–7923–0123–4
12. P. Fabbri (ed.): *Recreational Uses of Coastal Areas.* A Research Project of the Commission on the Coastal Environment, International Geographical Union. 1990 ISBN 0–7923–0279–6
13. L. M. Brush, M. G. Wolman and Huang Bing-Wei (eds.): *Taming the Yellow River: Silt and Floods.* Proceedings of a Bilateral Seminar on Problems in the Lower Reaches of the Yellow River, China. 1989 ISBN 0–7923–0416–0
14. J. Stillwell and H. J. Scholten (eds.): *Contemporary Research in Population Geography.* A Comparison of the United Kingdom and the Netherlands. 1990
    ISBN 0–7923–0431–4
15. M. S. Kenzer (ed.): *Applied Geography.* Issues, Questions, and Concerns. 1989 ISBN 0–7923–0438–1
16. D. Nir: *Region as a Socio-environmental System.* An Introduction to a Systemic Regional Geography. 1990 ISBN 0–7923–0516–7